STATISTICS IN PLAIN ENGLISH

STATISTICS IN PLAIN ENGLISH

Timothy C. Urdan
Santa Clara University

2001

LAWRENCE ERLBAUM ASSOCIATION, PUBLISHERS
Mahwah, New Jersey   London

Lawrence Erlbaum Associates, Inc., Publishers
10 Industrial Avenue
Mahwah, New Jersey 07430

Cover design by Kathryn Houghtaling Lacey

Library of Congress Cataloging-in-Publication Data

Urdan, Tim
   Statistics in plain English / by Tim Urdan.
     p. cm.
   Includes bibliographical references and index.
   ISBN 0-8058-3442-7 (pbk. : alk. paper)
    I. Statistics. 1. Title.

QA276. 12 U75 2001
519.5—dc21

                        00-061755

Books published by Lawrence Erlbaum Associates are printed on acid-free paper, and their bindings are chosen for strength and durability.

Printed in the United States of America
10   9   8   7   6   5   4   3   2   1

For Jeannine and Ella

CONTENTS

PREFACE

Why Use Statistics?

As a researcher who uses statistics frequently, and as an avid listener of talk radio, I find myself yelling at my radio daily. Although I realize that my cries go unheard, I cannot help myself. As radio talk show hosts, politicians making political speeches, and the general public all know, there is nothing more powerful and persuasive than the personal story, or what statisticians call anecdotal evidence. My favorite example of this comes from an exchange I had with a staff member of my congressman some years ago. I called his office to complain about a pamphlet his office had sent to me decrying the pathetic state of public education. I spoke to his staff member in charge of education. I told her, using statistics reported in a variety of sources (e.g., Berliner and Biddle's *The Manufactured Crisis* and the annual "Condition of Education" reports in the *Phi Delta Kappan* written by Gerald Bracey), that there are many signs that our system is doing quite well, including higher graduation rates, greater numbers of students in college, rising standardized test scores, and modest gains in SAT scores for all races of students. The staff member told me that despite these statistics, she knew our public schools were failing because she attended the same high school her father had, and he received a better education than she. I hung up and yelled at my phone.

Many people have a general distrust of statistics, believing that crafty statisticians can "make statistics say whatever they want" or "lie with statistics." In fact, if a researcher calculates the statistics correctly, he or she cannot make them say anything other than what they say, and statistics never lie. Rather, crafty researchers can interpret what the statistics *mean* in a variety of ways, and those who do not understand statistics are forced to either accept the interpretations that statisticians and researchers offer or to reject statistics completely. I believe a better option is to gain an understanding of how statistics work and then use that understanding to interpret the statistics one sees and hears for oneself. The purpose of this book is to make it a little easier to understand statistics.

Uses of Statistics

One of the potential shortfalls of anecdotal data is that they are idiosyncratic. Just as the congressional staffer told me her father received a better education from the high school they both attended than she did, I could have easily received a higher quality education than my father did. Statistics allow researchers to collect information, or data, from a large number of people and then summarize their typical experience. Do *most* people receive a better or worse education than their parents? Statistics allow researchers to take a large batch of data and *summarize* it into a couple of numbers, such as an average. Of course, when a large amount of data is summarized into a single number, a lot of information is lost, including the fact that different people have very different experiences. So it is important to remember that, for the most part, statistics do not provide useful information about each individual's experience. Rather, researchers generally use statistics to make *general* statements about a population. Although personal stories are often moving or interesting, it is often important to understand what the *typical* or *average* experience is. For this, we need statistics.

Statistics are also used to reach conclusions about general differences between groups. For example, suppose that in my family, there are four children, two men and two women. Suppose that the women in my family are taller than the men. This personal experience may lead me to the conclusion that women are generally taller than men. Of course, we know that, on average, men are taller than women. The reason we know this is because researchers have taken large, random samples of men and women and compared their average heights. Researchers are often interested in making such comparisons: Do cancer patients survive longer using one drug than another? Is one method of teaching children to read more effective than another? Do men and women differ in their enjoyment of a certain movie? To answer these questions, we need to collect data from randomly selected samples and compare these data using statistics. The results we get from such comparisons are often more trustworthy than the simple observations people make from non-random samples, such as the different heights of men and women in my family.

Statistics can also be used to see if scores on two variables are related and to make predictions. For example, statistics can be used to see whether smoking cigarettes is related to the likelihood of developing lung cancer. For years, tobacco companies argued that there was no relationship between smoking and cancer. Sure, some people who smoked developed cancer. But the tobacco companies argued that (a) many people who smoked never developed cancer, and (b) many people who smoke tend to do other things that may lead to cancer development, such as eating unhealthy foods and not exercising. With the help of statistics in a number of studies, researchers were finally able to produce a preponderance of evidence indicating that, in fact, there is a relationship between cigarette smoking and cancer. Because statistics tend to focus on overall patterns rather than individual cases, this research did not suggest that *everyone* who smokes would develop cancer. Rather, the research demonstrated that, on average, people have a greater chance of developing cancer if they smoke cigarettes than if they do not.

With a moment's thought, you can imagine a large number of interesting and important questions that statistics about relationships can help you answer. Is there a relationship between self-esteem and academic achievement? Is there a relationship between the appearance of criminal defendants and their likelihood of being convicted? Is it possible to predict the violent crime rate of a state from the amount of money the state spends on drug treatment programs? If we know the fathers' height, how accurately can we predict sons' height? These and thousands of other questions have been examined by researchers using statistics designed to determine the relationship between variables in a population.

How to Use This Book

This book is not intended to be used as a primary source of information for those who are unfamiliar with statistics. Rather, it is meant to be a supplement to a more detailed statistics textbook, such as that recommended for a statistics course in the social sciences. Or, if you have already taken a course or two in statistics, this book may be useful as a reference book to refresh your memory about statistical concepts you have encountered in the past. It is important to remember that this book is much less detailed than a traditional textbook. Each of the concepts discussed in this book is more complex than the presentation in this book would suggest, and a thorough understanding of these concepts may be acquired only with the use of a more traditional, more detailed textbook.

With that warning firmly in mind, let me describe the potential benefits of this book, and how to make the most of them. As a researcher and a teacher of statistics, I have found that statistics textbooks often contain a lot of technical information that can be intimidating to non-statisticians. Although, as I said previously, this information is important, sometimes it is useful to have a short, simple description of a statistic, when it should be used, and how to make sense of it. This is particularly true for students taking only their first or second statistics course, those who do not consider themselves to be "mathematically inclined," and those who may have taken statistics years ago and now find themselves in need of a little refresher. My purpose in writing this book is to provide short, simple descriptions and explanations of a number of statistics that are easy to read and understand.

To help you use this book in a manner that best suits your needs, I have organized each chapter into three sections. In the first section, a brief (one to two pages) description of the statistic is given, including what the statistic is used for and what information it provides. The second section of each chapter contains a slightly longer (three to eight pages) discussion of the statistic. In this section, I provide a bit more information about how the statistic works, an explanation of how the formula for calculating the statistic works, the strengths and weaknesses of the statistic, and the conditions that must exist to use the statistic. Finally, each chapter concludes with an example in which the statistic is used and interpreted.

Before reading the book, it may be helpful to note three of its features. First, some of the chapters discuss more than one statistic. For example, in Chapter 1, three measures of central tendency are described: the mean, median, and mode. Second, some of the chapters cover statistical concepts rather than specific statistical techniques. For example, in Chapter 3 the normal distribution is discussed. There are also chapters on statistical significance and on statistical interactions.   Finally, you should remember that the chapters in this book are not necessarily designed to be read in order. The book is organized such that the more basic statistics and statistical concepts are in the earlier chapters whereas the more complex concepts appear later in the book. However, it is not necessary to read one chapter before understanding the next. Rather, each chapter in the book was written to stand on its own. This was done so that you could use each chapter as needed. If, for example, you had no problem understanding t tests when you learned about them in your statistics class but find yourself struggling to understand one-way analysis of variance, you may want to skip the t test chapter (Chapter 8) and skip directly to the analysis of variance chapter (Chapter 9).

Statistics are powerful tools that help people understand interesting phenomena. Whether you are a student, a researcher, or just a citizen interested in understanding the world around you, statistics can offer one method for helping you make sense of your environment. This book was written using plain English to make it easier for non-statisticians to take advantage of the many benefits statistics can offer. I hope you find it useful.

CHAPTER 1

MEASURES OF CENTRAL TENDENCY

Whenever you collect data, you end up with a group of scores on one or more variables. If you take the scores on one variable and arrange them in order from lowest to highest, what you get is a **distribution** of scores. Researchers often want to know about the characteristics of these distributions of scores, such as the shape of the distribution, how spread out the scores are, what the most common score is, and so on. One set of distribution characteristics that researchers are usually interested in is central tendency. This set consists of the mean, median, and mode.

The **mean** is probably the most commonly used statistic in all social science research. The mean is simply the arithmetic average of a distribution of scores, and researchers like it because it provides a single, simple number that gives a rough summary of the distribution. It is important to remember that although the mean provides a useful piece of information, it does not tell you anything about how spread out the scores are (i.e., variance) or how many scores in the distribution are close to the mean. It is possible for a distribution to have very few scores at or near the mean.

The **median** is the score in the distribution that marks the 50th percentile. That is, 50% percent of the scores in the distribution fall above the median and 50% fall below it. Researchers often use the median when they want to divide their distribution scores into two equal groups (called a **median split**). The median is also a useful statistic to examine when the scores in a distribution are skewed or when there are a few extreme scores at the high end or the low end of the distribution. This is discussed in more detail in the following pages.

The **mode** is the least used of the measures of central tendency because it provides the least amount of information. The mode simply indicates which score in the distribution occurs most often, or has the highest frequency.

A Word About Populations and Samples

You will notice in Table 1.1 that there are two different symbols used for the mean, $\overline{X}$ and μ. Two different symbols are needed because it is important to distinguish between a **statistic** that applies to a **sample** and a **parameter** that applies to a **population**. A population is an individual or group that represents all the members of a certain group or category of interest. A sample is a subset drawn from the larger population. For example, suppose that I wanted to know the average income of the current full-time, tenured faculty at Harvard. There are two ways that I could find this average. First, I could get a list of every full-time, tenured faculty member at Harvard and find out the annual income of each member on this list. Because this list contains every member of the group that I am interested in, it can be considered a population. If I were to collect these data and calculate the mean, I would have generated a parameter, because a parameter is a value generated from, or applied to, a population. The symbol used to represent the population mean is μ. Another way to generate the mean income of the tenured faculty at Harvard would be to randomly select a sub-set of faculty names from my list and calculate the average income of this sub-set. The sub-set is known as a sample (in this case it is a random sample), and the mean that I generate from this sample is a type of statistic. Statistics are values derived from sample data, whereas parameters are values that are either derived from, or applied to, population data. It is important to note that all samples are representative of some population and that all sample statistics can be used as estimates of population parameters. In the case of the mean, the sample statistic is represented with the symbol $\overline{X}$. The distinction between sample statistics and population parameters appears again in later chapters (e.g., Chapters 3 and 6).

MEASURES OF CENTRAL TENDENCY IN DEPTH

The calculations for each measure of central tendency are mercifully straightforward. With the aid of a calculator or statistics software program, you will probably never need to calculate any of these statistics. But for the sake of knowledge and in the event you find yourself without a calculator and in need of these statistics, here is the information you will need.

Because the mean is an average, calculating the mean involves adding, or summing, all of the scores in a distribution and dividing by the number of scores. So, if you have 10 scores in a distribution, you would add all of the scores together to find the sum and then divide the sum by 10, which is the number of scores in the distribution. The formula for calculating the mean is presented in Table 1.1.

TABLE 1.1 Formula for calculating the mean of a distribution.

$$\overline{X}, \mu = \frac{\Sigma X}{n, N}$$

where $\overline{X}$ is the sample mean
μ is the population mean
Σ means "the sum of"
X is an individual score in the distribution
n is the number of scores in the sample
N is the number of scores in the population

The calculation of the median for a simple distribution of scores[1] is even simpler than the calculation of the mean. To find the median of a distribution, you need to first arrange all of the scores in the distribution in order, from smallest to largest. Once this is done, you simply need to find the middle score in the distribution. If there are an odd number of scores in the distribution, there will be a single score that marks the middle of the distribution. For example, if there are 11 scores in the distribution arranged in descending order from smallest to largest, the 6th score will be the median because there will be 5 scores below it and 5 scores above it. However, if there are an even number of scores in the distribution, there is no single middle score. In this case, the median is the average of the two scores in the middle of the distribution (as long as the scores are arranged in order, from largest to smallest). For example, if there are 10 scores in a distribution, to find the median you will need to find the average of the 5th and 6th scores. To find this average, you simply add the two scores together and divide by two.

To find the mode, there is no need to calculate anything. The mode is simply the category in the distribution that has the highest number of scores, or the highest frequency. For example, suppose you have the following distribution of IQ test scores from 10 students:

86   90   95   100   100   100   110   110   115   120

In this distribution, the score that occurs most frequently is **100**, making it the mode of the distribution. If a distribution has more than one category with the most common score, the distribution has multiple modes and is called **multimodal**. One common example of a multimodal distribution is the **bimodal** distribution. Researchers often get bimodal distributions when they ask people to respond to controversial questions that tend to polarize the public. For example, if I were to ask a sample of 100 people how they feel about capital punishment, I might get the results presented in Table 1.2.

---

[1] It is also possible to calculate the median of a **grouped frequency distribution.** For an excellent description of the technique for calculating a median from a grouped frequency distribution, see Spatz (2001) *Basic Statistics: Tales of Distributions* (7th ed.).

On the following scale, please indicate how you feel about capital punishment.

1----------2----------3----------4----------5
Strongly   Strongly
Opposed   in Favor

TABLE 1.2   Frequency of responses.

	Category of Responses on the Scale				
	1	*2*	*3*	*4*	*5*
Frequency of Responses in Each Category	45	3	4	3	45

In this example, because most people either strongly opposed or strongly supported capital punishment, I ended up with a bimodal distribution of scores.

EXAMPLE: THE MEAN, MEDIAN, AND MODE OF A SKEWED DISTRIBUTION

As you will see in Chapter 3, when scores in a distribution are normally distributed, the mean, median, and mode are all at the same point: the center of the distribution. In the messy world of social science, however, the scores from a sample on a given variable are often not normally distributed. When the scores in a distribution tend to bunch up at one end of the distribution and there are a few scores at the other end, the distribution is said to be skewed. When working with a skewed distribution, the mean, median, and mode are usually all at different points.

It is important to note that the procedures used to calculate a mean, median, and mode are the same whether you are dealing with a skewed or a normal distribution. All that changes are where these three measures of central tendency are in relation to each other. To illustrate, I created a fictional distribution of scores based on a sample size of 30. Suppose that I were to ask a sample of 30 randomly selected fifth graders whether they think it is important to do well in school. Suppose, further, that I ask them to rate how important they think it is to do well in school using a 5-point scale, with 1= "not at all important" and 5 = "very important." Because most fifth graders tend to believe it is very important to do well in school, most of the scores in this distribution are at the high end of the scale, with a few scores at the low end. I have arranged my fictitious scores in order from smallest to largest and get the following distribution:

1	1	1	2	2	2	3	3	3	3
4	4	4	4	4	4	4	4	5	5
5	5	5	5	5	5	5	5	5	5

As you can see, there are only a few scores near the low end of the distribution (1's and 2's) and more at the high end of the distribution (4's and 5's). To get a clear picture of what this skewed distribution looks like, I have created the graph in Figure 1.1.

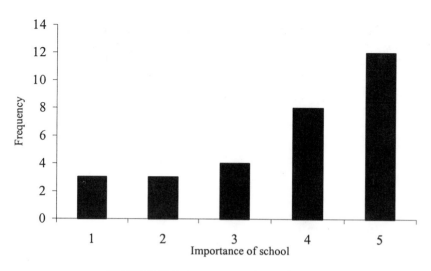

FIGURE 1.1 A skewed distribution.

This graph provides a picture of what some skewed distributions look like. Notice how most of the scores are clustered at the higher end of the distribution and there are a few scores creating a tail toward the lower end. This is known as a **negatively skewed** distribution, because the tail goes toward the lower end. If the tail of the distribution were pulled out toward the higher end, this would have been a **positively skewed** distribution.

A quick glance at the scores in the distribution, or at the graph, reveals that the mode is 5 because there were more scores of 5 than any other number in the distribution.

To calculate the mean, we simply apply the formula mentioned earlier. That is, we add up all of the scores (ΣX) and then divide this sum by the number of scores in the distribution (n). This gives us a fraction of 113/30, which reduces to 3.7666. When we round to the second place after the decimal, we end up with a mean of 3.77.

To find the median of this distribution, we arrange the scores in order from smallest to largest and find the middle score. In this distribution, there are 30 scores, so there will be 2 in the middle. When arranged in order, the 2 scores in the middle (the 15th and 16th scores) are both 4. When we add these two scores together and divide by 2, we end up with 4, making our median 4.

As I mentioned earlier, the mean of a distribution can be affected by scores that are unusually large or small for a distribution, sometimes called **outliers**, whereas the median is not affected by such scores. In the case of a skewed distribution, the mean is usually pulled in the direction of the tail, because the tail is where the outliers are. In a negatively skewed distribution, such as the one presented previously, we would expect the mean to be smaller than the median, because the mean is pulled toward the tail whereas the median is not. In our example, the mean (3.77) is somewhat lower than the median (4). In positively skewed distributions, the mean is usually somewhat higher than the median.

WRAPPING UP AND LOOKING FORWARD

Measures of central tendency, particularly the mean and the median, are some of the most used and useful statistics for researchers. They each provide important information about an entire distribution of scores in a single number. For example, we know that the average height of a man in the United States is five feet nine inches tall. This single number is used to summarize information about millions of men in this country. But for the same reason that the mean and median are useful, they can often be dangerous if we forget that a statistic such as the mean *ignores* a lot of information about a distribution, including the great amount of variety that exists in many distributions. Without considering the variety as well as the average, it becomes easy to make sweeping generalizations, or stereotypes, based on the mean. The measure of variance is the topic of the next chapter.

GLOSSARY OF TERMS AND SYMBOLS FOR CHAPTER 1

Bimodal: A distribution that has two values that have the highest frequency of scores.

Distribution: A collection, or group, of scores from a sample on a single variable. Often, but not necessarily, these scores are arranged in order from smallest to largest.

Mean: The arithmetic average of a distribution of scores.

Median split: Dividing a distribution of scores into two equal groups by using the median score as the divider. Those scores above the median are the "high" group whereas those below the median are the "low" group.

Median: The score in a distribution that marks the 50th percentile. It is the score at which 50% of the distribution falls below and 50% fall above.

Mode: The score in the distribution that occurs most frequently.

Multimodal: When a distribution of scores has two or more values that have the highest frequency of scores.

Negative skew: In a skewed distribution, when most of the scores are clustered at the higher end of the distribution with a few scores creating a tail at the lower end of the distribution.

Outliers: Extreme scores that are more than two standard deviations above or below the mean.

Positive skew: In a skewed distribution, when most of the scores are clustered at the lower end of the distribution with a few scores creating a tail at the higher end of the distribution.

Parameter: A value derived from the data collected from a population, or the value inferred to the population from a sample statistic.

Population: The group from which data are collected or a sample is selected. The population encompasses the entire group for which the data are alleged to apply.

Sample: An individual or group, selected from a population, from whom or which data are collected.

Skew: When a distribution of scores has a high number of scores clustered at one end of the distribution with relatively few scores spread out toward the other end of the distribution, forming a tail.

Statistic: A value derived from the data collected from a sample.

Σ	The sum of; to sum.
X	An individual score in a distribution.
ΣX	The sum of X; adding up all of the scores in a distribution.
$\overline{X}$	The mean of a sample.
μ	The mean of a population.
n	The number of cases, or scores, in a sample.
N	The number of cases, or scores, in a population.

CHAPTER 2

MEASURES OF VARIABILITY

Measures of central tendency, such as the mean and the median described in Chapter 1, provide useful information. But it is important to recognize that these measures are limited and, by themselves, do not provide a great deal of information. To illustrate, consider this example: Suppose I gave a sample of 100 fifth-grade children a survey to assess their level of depression. Suppose further that this sample had a mean of 10.0 on my depression survey and a median of 10.0 as well. All we know from this information is that the mean and median are in the same place in my distribution, and this place is 10.0. Now consider what we do not know. We do not know if this is a high score or a low score. We do not know if all of the students in my sample have about the same level of depression or if they differ from each other. We do not know the highest depression score in our distribution or the lowest score. Simply put, we do not yet know anything about the dispersion of scores in the distribution. In other words, we do not yet know anything about the variety of the scores in the distribution.

There are three measures of dispersion that researchers typically examine: the **range**, the **variance**, and the **standard deviation**. Of these, the standard deviation is perhaps the most informative and certainly the most widely used.

Range

The range is simply the difference between the largest score (the **maximum value**) and the smallest (the **minimum value**) score of a distribution. This statistic gives researchers a quick sense of how spread out the scores of a distribution are, but it is not a particularly useful statistic because it can be quite misleading. For example, in our depression survey described earlier, we may have 1 student score a 1 and another score a 20, but the other 98 may all score 10. In this example, the range will be 19 (20 – 1 = 19), but the scores really are not as spread out as the range might suggest. Researchers often take a quick look at the range to see whether all or most of the points on a scale, such as a survey, were covered in the sample.

Another common measure of the range of scores in a distribution is the **interquartile range (IQR).** Unlike the range, which is the difference between the largest and smallest score in the distribution, the IQR is the difference between the score that marks the 75th percentile (the third quartile) and the score that marks the 25th percentile (the first quartile). If the scores in a distribution were arranged in order from largest to smallest and then divided into groups of equal size, the IQR would contain the scores in the two middle quartiles.

Variance

The variance provides a statistical average of the amount of dispersion in a distribution of scores. Because of the mathematical manipulation needed to produce a variance statistic (more about this in the next section), variance, by itself, it not often used by researchers to gain a sense of a distribution. In general, variance is used more as a step in the calculation of other statistics (e.g., analysis of variance) than as a stand-alone statistic. But with a simple manipulation, the variance can be transformed into the standard deviation, which is one of the statistician's favorite tools.

Standard Deviation

The best way to understand what a standard deviation is to consider what the two words mean. *Deviation*, in this case, refers to the difference between an individual score in a distribution and the average score for the distribution. So if the average score for a distribution is 10 (as in our previous example), and an individual child has a score of 12, the deviation is 2. The other word in the term standard deviation is *standard*. In this case, standard means typical, or average. So a standard deviation is the typical, or average, deviation between individual scores in a distribution and the mean for the distribution.[2] This is a very useful statistic because it provides a handy measure of how spread out the scores are in the distribution. When combined, the mean and standard deviation provide a pretty good picture of what the distribution of scores is like.

In a sense, the range provides a measure of the total spread in a distribution (i.e., from the lowest to the highest scores), whereas the variance and standard deviation are measures of the average amount of spread within the distribution. Researchers tend to look at the range when they want a quick snapshot of a distribution, such as when they want to know whether all of the response categories on a survey question have been used (i.e., did people use all 5 points on the 5-point Likert scale?) or they want a sense of the overall balance of scores in the distribution. Researchers rarely look at the variance alone, because it does not use the same scales as the original measure of a variable, although the variance statistic is very useful for the calculation of other statistics (such as analysis of variance, Chapter 9). The standard deviation is a very useful statistic that researchers constantly examine to provide the most easily interpretable and meaningful measure of the average dispersion of scores in a distribution.

MEASURES OF VARIABILITY IN DEPTH

Calculating the Variance and Standard Deviation

There are two central issues that I need to address when considering the formulas for calculating the variance and standard deviation of a distribution: (a) whether to use the formula for the sample or the population, and (b) how to make sense of these formulas.

Sample Statistics as Estimates of Population Parameters

Although researchers are sometimes interested in simply describing the characteristics (e.g., mean, median, range, etc.) of a sample, for the most part we are much more concerned with what our sample tells us about the population from which the sample was drawn. If I am conducting a study about the study habits of eighth graders, I am more concerned about eighth graders *in general* than I am about the particular group of eighth graders in my sample. It is important to keep this in mind, because most of our statistics, although generated from sample data, are used to make estimations about the population. This point is discussed in more detail later in the book when examining inferential statistics. But it is important to keep this in mind as you read about these measures of variation. The formulas for calculating the variance and standard deviation of sample data are actually designed to make these sample statistics better *estimates* of the population parameters (i.e., the population variance and standard deviation).

It is important to note that the formulas for calculating the variance and the standard deviation differ depending on whether you are working with a distribution of scores taken from a sample or from a population. The reason these two formulas are different is quite complex, and requires more space than allowed in a short book like this. I provide an overly brief explanation here and then encourage you to find a more thorough explanation in a traditional statistics textbook. Briefly, when we do not know the population mean, we must use the sample mean as an estimate. But the sample mean will probably differ from the population mean. Whenever we use a number *other than* the actual mean to calculate the variance, we will end up with a *larger* variance, and therefore a larger standard deviation, than if we had used the actual mean. This will be true

---

[2] Although the standard deviation is technically not the "average deviation" for a distribution of scores, in practice this is a useful heuristic for gaining a rough conceptual understanding of what this statistic is. The actual formula for the average deviation would be $\Sigma(|X - \text{mean}|)/N$.

regardless of whether the number we use in our formula is smaller or larger than our actual mean. Because the sample mean usually differs from the population mean, the variance and standard deviation that we calculate using the sample mean will probably be smaller than it would have been had we used the population mean. Therefore, when we use the sample mean to generate an *estimate* of the population variance or standard deviation, we will actually *under*estimate the size of the population mean. To adjust for this underestimation, we use $n - 1$ in the denominator of our sample formulas. Smaller denominators produce larger overall variance and standard deviation statistics, which will be more accurate estimates of the population parameters.

The formulas for calculating the variance and standard deviation of a population and the estimates of the population variance and standard deviation based on a sample are presented in Table 2.1. As you can see, the formulas for calculating the variance and the standard deviation are virtually identical. Because both require that you calculate the variance first, we begin with the formulas for calculating the variance (see Table 2.1). This formula is known as the *deviation score formula.*[3]

When working with a population distribution, the formulas for the variance and for the standard deviation both have a denominator of N, which is the size of the population. In the real world of research, particularly social science research, we usually assume that we are working with a sample that represents a larger population. For example, if I study the effectiveness of my new reading program with a class of second graders, as a researcher I assume that these particular second graders represent a larger population of second graders, or students more generally. Because of this type of inference, researchers generally think of their research participants as a sample rather than a population, and the formula for calculating the variance of a sample is the more often used. Notice that the formula for calculating the variance of a sample is identical to that used for the population except the denominator for the sample formula is $n - 1$.

How much of a difference does it make if we use N or $n - 1$ in our denominator? Well, that depends on the size of the sample. If we have a sample of 500 people, there is virtually no difference between the variance formula for the population and for the estimate based on the sample. After all, dividing a numerator by 500 is almost the same as dividing it by 499. But when we have a small sample, such as a sample of 10, then there is a relatively large difference between the results produced by the population and sample formulas.

The second issue to address involves making sense of the formulas for calculating the variance. In all honesty, there will be very few times that you will need to use this formula. Outside of my teaching duties, I haven't calculated a standard deviation by hand since my first statistics course. Thankfully, all computer statistics and spreadsheet programs, and many calculators, compute the variance and standard deviation for us. Nevertheless, it is mildly interesting and quite informative to examine how these variance formulas work.

To begin this examination, let me remind you that the variance is simply an average of a distribution. To get an average, we need to add up all of the scores in a distribution and divide this sum by the number of scores in the distribution, which is n (remember the formula for calculating the mean in Chapter 1?). With the variance, however, we need to remember that we are not interested in the average *score* of the distribution. Rather, we are interested in the average *difference,* or *deviation,* between each score in the distribution and the mean of the distribution. To get this information, we have to calculate a *deviation score* for each individual score in the distribution. This score is calculated by taking an individual score and subtracting the mean from that score. If we compute a deviation score for each individual score in the distribution, then we can add these deviation scores up and divide by n to get the average, or standard, deviation, right? Not quite.

The problem here is that, by definition, the mean of a distribution is the mathematical middle of the distribution. Therefore, some of the scores in the distribution will fall above the mean (producing positive deviation scores), and some will fall below the mean (producing negative deviation scores). When we add these positive and negative deviation scores together, the sum will be zero. Because the mean is the mathematical middle of the distribution, we will get zero when we add up these deviation scores no matter how big or small our sample, or how skewed or normal our

---

[3] It is also possible to calculate the variance and standard deviation using the *raw score formula*, which does not require that you calculate the mean. The raw score formula is included in most standard statistics textbooks.

distribution. And because we cannot find an average of zero (i.e., zero divided by n is zero, no matter what n is), we need to do something to get rid of this zero.

TABLE 2.1 Variance and standard deviation formulas.

	Population	*Estimate Based on a Sample*
Variance	$\sigma^2 = \dfrac{\Sigma(X - \mu)^2}{N}$	$s^2 = \dfrac{\Sigma(X - \overline{X})^2}{n-1}$
	where Σ = to sum X = a score in the distribution μ = the population mean N = the number of cases in the population	where Σ = to sum X = a score in the distribution $\overline{X}$ = the sample mean n = the number of cases in the sample
Standard Deviation	$\sigma = \sqrt{\dfrac{\Sigma(X - \mu)^2}{N}}$	$s = \sqrt{\dfrac{\Sigma(X - \overline{X})^2}{n-1}}$
	where Σ = to sum X = a score in the distribution μ = the population mean N = the number of cases in the population	where Σ = to sum X = a score in the distribution $\overline{X}$ = the sample mean n = the number of cases in the sample

The solution statisticians came up with is to make each deviation score positive by squaring it. So, for each score in a distribution, we subtract the mean of the distribution and then square the deviation. If you look at the deviation score formulas in Table 2.1, you will see that all that the formula is doing with $(X - \mu)^2$ is to take each score, subtract the mean, and square the resulting deviation score. What you get when you do this is the all-important **squared deviation**, which is used all the time in statistics. If we then put a summation sign in front such that we have $\Sigma(X - \mu)^2$. What this tells us is that after we have produced a squared deviation score for each case in our distribution, we then need to add up all of these squared deviations, giving us the **sum of squared deviations**, or the **sum of squares (SS)**. Once this is done, we divide by the number of cases in our distribution, and we get an average, or mean, of the squared deviations. This is our variance.

The final step in this process is converting the variance into a standard deviation. Remember that in order to calculate the variance, we had to square each deviation score. We did this to avoid getting a sum of zero in our numerator. When we squared these scores, we changed our statistic from our original scale of measurement (i.e., whatever units of measurement were used to generate our distribution of scores) to a squared score. To reverse this process and give us a statistic that is back to our original unit of measurement, we merely need to take the square root of our variance. When we do this, we switch from the variance to the standard deviation. Therefore, the formula for calculating the standard deviation is exactly the same as the formula for calculating the variance, except we put a big square root symbol over the whole formula. Notice that because of the squaring and square rooting process, the standard deviation and the variance are always positive numbers.

Why Have Variance?

If the variance is a difficult statistic to understand, and rarely examined by researchers, why not just eliminate this statistic and jump straight to the standard deviation? There are two reasons. First, we need to calculate the variance before we can find the standard deviation anyway, so it is not more work. Second, the fundamental piece of the variance formula, which is the sum of the squared deviations, is used in a number of other statistics, most notably analysis of variance (ANOVA). When you learn about more advanced statistics such as ANOVA, factorial ANOVA, and even regression, you will see that each of these uses the *sum of squares,* which is just another way of saying the sum of the squared deviations. Because the sum of squares is such an important piece of so many statistics, the variance statistic has maintained a place in the teaching of basic statistics.

EXAMPLE: EXAMINING THE RANGE, VARIANCE, AND STANDARD DEVIATION

I recently conducted a study in which I gave questionnaires to approximately 500 high school students in the 9th and 11th grades. In the examples that follow, we examine the mean, range, variance, and standard deviation of the distribution of responses to two of these questions. To make sense of these (and all) statistics, you need to know exactly what the wording of the survey items was and the response scale that the students in the sample used to answer the survey items. Although this may sound obvious, I mention it here because, if you notice, much of the statistical information reported in the news (e.g., the results of polls) does not provide the exact wording of the questions or the choices the sample had to respond to the questions. Without this information, it is difficult to know exactly what the responses mean, and "lying with statistics" becomes easier.

The first survey item we examine reads "If I have enough time, I can do even the most difficult work in this class." This item is designed to measure students' confidence in their abilities to succeed in their classwork. Students were asked to respond to this question by circling a number on a scale from 1 to 5. On this scale, circling the 1 means that the statement is "not at all true" and the 5 means "very true." So students were basically asked to indicate how true they felt the statement was on a scale from 1 to 5, with higher numbers indicating a stronger belief that the statement was true.

I received responses from 491 students on this item. The distribution of responses produced the following statistics:

$$\text{Sample Size} = 491$$
$$\text{Mean} = 4.21$$
$$\text{Standard Deviation} = .98$$
$$\text{Variance} = (.98)^2 = .96$$
$$\text{Range} = 5 - 1 = 4$$

A graph of the frequency distribution for the responses on this item appears in Figure 2.1. As you can see in this graph, most of the students in the sample circled number 4 or number 5 on the response scale, indicating that they felt the item was quite true (i.e., that they were confident in their ability to do their classwork if they were given enough time). Because most students circled a 4 or a 5, the average score on this item is quite high (4.21 out of a possible 5). This is a negatively skewed distribution.

The graph in Figure 2.1 also provides information about the variety of scores in this distribution. Although our range statistic is 4, indicating that students in the sample circled both the highest and the lowest number on the response scale, we can see that the range does not really provide much useful information. For example, the range does not tell us that most of the students in our sample scored at the high end of the scale. By combining the information from the range statistic with the mean statistic, we can reach the following conclusion: "Although the distribution of scores on this item covers the full range, it appears that most scores are at the higher end of the response scale."

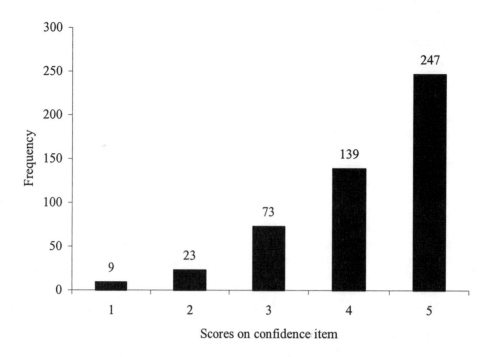

FIGURE 2.1 Frequency distribution of scores on confidence item.

Now that we've determined that (a) the distribution of scores covers the full range of possible scores (i.e., from 1 to 5), and (b) most of the responses were at the high end of the scale (because the mean is 4.21 out of a possible 5), we may want a more precise measure of the average amount of variety among the scores in the distribution. For this we turn to the variance and standard deviation statistics. In this example, the variance (.96) is almost exactly the same as the standard deviation (.98). This is something of a fluke. Do not be fooled. It is quite rare for the variance and standard deviation to be so similar. In fact, this only happens if the standard deviation is about 1.0, because 1.0 squared is 1.0. So in this rare case, the variance and standard deviation provide almost the same information. Namely, they indicate that the average difference between an individual score in the distribution and the mean for the distribution is about 1 point on the 5-point scale.

Taken together, these statistics tell us the same things that the graph tells us, but more precisely. Namely, we now know that (a) students in the study answered this item covering the whole range of response choices (i.e., 1 – 5); (b) most of the students answered at or near the top of the range, because the mean is quite high; and (c) the scores in this distribution generally packed fairly closely together with most students circling a number within 1 point of the mean, because the standard deviation was .98. The variance tells us that the average *squared deviation* was .96, and we scratch our heads, wonder what good it does us to know the average squared deviation, and move on.

In our second example, we examine students' responses to the item "I would feel really good if I were the only one who could answer the teacher's question in class." This item is one of several on the survey designed to measure students' desires to demonstrate to others that they are smart, or academically able.

We received responses from 491 students on this item and the distribution produced the following statistics:

$$\text{Sample Size} = 491$$
$$\text{Mean} = 2.92$$
$$\text{Standard Deviation} = 1.43$$
$$\text{Variance} = (1.43)^2 = 2.04$$
$$\text{Range} = 5 - 1 = 4$$

In Figure 2.2, a graph is presented that illustrates the distribution of students' responses to this item across each of the five response categories. It is obvious, when looking at this graph, how the distribution of scores on this item differs from the distribution of scores on the confidence item presented earlier. But if we didn't have this graph, how could we use the statistics to discover these differences between the distributions of scores on these two items?

Notice that, as with the previous item, the range is 4, indicating that some students circled the number 1 on the response scale and some circled the number 5. Because the ranges for both the confidence and the wanting to appear able items are equal (i.e., 4), they do nothing to indicate the differences in the distributions of the responses to these two items. That is why the range is not a particularly useful statistic—it simply does not provide very much information.

Our first real indication that the distributions differ substantially comes from a comparison of the means. In the previous example, the mean of 4.21 indicated that most of the students must have circled either a 4 or a 5 on the response scale. For this second item, the mean of 2.92 is a bit less informative. Although it provides an average score, it is impossible from just examining the mean to determine whether most students circled a 2 or 3 on the scale, or whether roughly equal numbers of students circled each of the five numbers on the response scale, or whether almost half of the students circled 1 whereas the other half circled 5. All three scenarios would produce a mean of about 2.92, because that is roughly the middle of the response scale.

To get a better picture of this distribution, we need to consider the standard deviation in conjunction with the mean. Before discussing the actual standard deviation for this distribution of scores, let us briefly consider what we would expect the standard deviation to be for each of the three scenarios just described. First, if almost all of the students circled a 2 or a 3 on the response scale, we would expect a fairly small standard deviation, as we saw in the previous example using the confidence item. The more similar the responses are to an item, the smaller the standard deviation. However, if half of the students circled 1 whereas the other half circled 5, we would expect a large standard deviation (about 2.0) because each score would be about two units away from the mean (if the mean is about 3.0 and each response is either 1 or 5, each response is about two units away from the mean). Finally, if the responses are fairly evenly spread out across the five response categories, we would expect a moderately sized standard deviation (about 1.50).

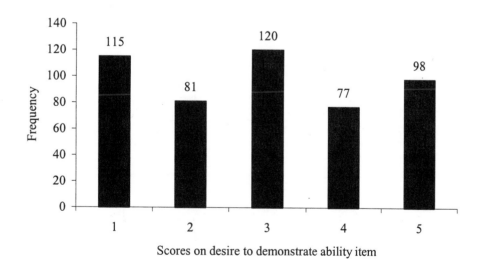

FIGURE 2.2 Frequency distribution for scores on desire to demonstrate ability item.

Now, when we look at the actual mean for this distribution (2.92) and the actual standard deviation (1.43), we can develop a rough picture of the distribution in our minds. Because we know that on a scale from 1 to 5, a mean of 2.92 is about in the middle, we can guess that the distribution looks somewhat symmetrical (i.e., that there will be roughly the same number of responses in the 4 and 5 categories as there are in the 1 and 2 categories. Furthermore, because we've got a moderately

sized standard deviation of 1.43, we know that the scores are pretty well spread out, with a healthy number of students in each of the five response categories. So we know
that we didn't get an overwhelming number of students circling 3 and we didn't get students circling only 1 or 5. At this point, that is about all we can say about this distribution: The mean is near the middle of the scale and the responses are pretty well spread out across the five response categories. To say any more, we would need to look at the number of responses in each category, such as that presented in Figure 2.2.

As we look at the actual distribution of scores presented in the graph in Figure 2.2, we can see that our predictions that we generated from our statistics about the shape of the distribution were pretty accurate. Notice that we did not need to consider the variance at all, because the variance in this example (2.04) is on a different scale of measurement than our original 5-point response scale, and therefore is very difficult to interpret. Variance is an important statistic for many techniques (e.g., ANOVA, regression), but it does little to help us understand the shape of a distribution of scores. The mean, standard deviation, and to a lesser extent the range, when considered together, can provide a rough picture of a distribution of scores. Often, a rough picture is all a researcher needs or wants. Sometimes, however, researchers need to know more precisely the characteristics of a distribution of scores. In that case, a picture, such as a graph, may be worth a thousand words.

Another useful way to examine a distribution of scores is to create a **boxplot**. In Figure 2.3, a boxplot is presented for the same variable that is represented in Figure 2.2, wanting to demonstrate ability. This boxplot was produced in the SPSS statistical software program. The box in this graph contains some very useful information. First, the thick line in the middle of the box represents the median of this distribution of scores. The top line of the box represents the 75th percentile of the distribution and the bottom line represents the 25th percentile. Therefore, the top and bottom lines of the box reveal the interquartile range (IQR) for this distribution. In other words, 50 percent of the scores on this variable in this distribution are contained within the upper and lower lines of this box (i.e., 50% of the scores are between just above a score of 2 and just below a score of 4). The vertical lines coming out of the top and bottom of the box and culminating in horizontal lines reveal the largest and smallest scores in the distribution, or the range. These scores are 5 and 1, producing a range of $5 - 1 = 4$. As you can see, the boxplot in Figure 2.3 contains a lot of useful information about the spread of scores on this variable in a single picture.

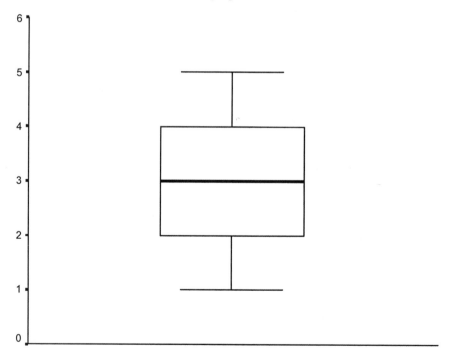

FIGURE 2.3 Boxplot for the desire to appear able variable.

WRAPPING UP AND LOOKING FORWARD

Measures of variation, such as the variance, standard deviation, and range, are important descriptive statistics. They provide useful information about how spread out the scores of a distribution are, and the shape of the distribution. Perhaps even more important than their utility as descriptors of a single distribution of scores is their role in more advanced statistics such as those coming in later chapters (e.g., ANOVA in Chapters 9, 10, and 11). In the next chapter, we examine the properties of a theoretical distribution, the normal distribution, that has a specific shape and characteristics. Using some of the concepts from Chapter 2, we can see how the normal distribution can be used make inferences about the population based on sample data.

GLOSSARY OF TERMS AND SYMBOLS FOR CHAPTER 2

Boxplot: A graphic representation of the distribution of scores on a variable that includes the range, the median, and the interquartile range.

Interquartile Range (IQR): The difference between the 75th percentile and 25th percentile scores in a distribution.

Range: The range is the difference between the largest score and the smallest score of a distribution.

Squared deviation: The difference between an individual score in a distribution and the mean for the distribution, squared.

Standard deviation: The average deviation between the individual scores in the distribution and the mean for the distribution.

Sum of squared deviations, sum of squares: The sum of each squared deviation for all of the cases in the sample.

Variance: The sum of the squared deviations divided by the number of cases in the population, or by the number of cases minus one in the sample.

μ	Symbol for the population mean.
X	Symbol for an individual score in a distribution.
s^2	Symbol for the sample variance.
s	Symbol for the sample standard deviation.
σ	Symbol for the population standard deviation.
σ^2	Symbol for the population variance.
SS	Symbol for the sum of squares, or sum of squared deviations.
n	Symbol for the number of cases in the sample.
N	Symbol for the number of cases in the population.

CHAPTER 3

THE NORMAL DISTRIBUTION

The **normal distribution** is a concept with which most people have some familiarity, although they often have never heard of the term. A more familiar name for the normal distribution is the **bell curve**, because a normal distribution forms the shape of a bell. The normal distribution is extremely important to statistics, and has some specific characteristics that make it so useful. In this chapter I begin by briefly describing what a normal distribution is and why it is so important to researchers. Then I discuss some of the features of the normal distribution, and of sampling, in more depth.

Characteristics of the Normal Distribution

If you take a look at the normal distribution shape presented in Figure 3.1, you may notice that the normal distribution has three fundamental characteristics. First, it is **symmetrical**, meaning that the upper half and the lower half of the distribution are mirror images of each other. Second, the mean, median, and mode are all in the same place, in the center of the distribution (i.e., the top of the bell curve). Because of this second feature, the normal distribution is highest in the middle, it is **unimodal**, and it curves downward toward the top and bottom of the distribution. Finally, the normal distribution is **asymptotic**, meaning that the upper and lower tails of the distribution never actually touch the baseline, also known as the X axis.

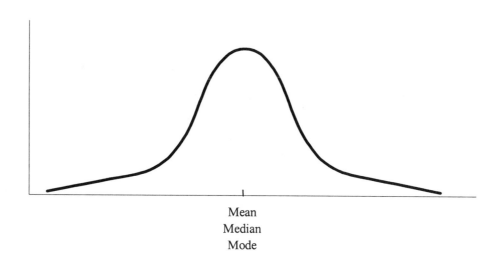

Mean
Median
Mode

FIGURE 3.1 The normal distribution.

Why Is the Normal Distribution So Important?

These three characteristics of the normal distribution are each critical in statistics because they allow us to make good use of **probability** statistics. When researchers collect data from a sample, sometimes all they want to know about are characteristics of the sample. For example, if I wanted to examine the eating habits of 100 first-year college students, I would just select 100 students, ask them what they eat, and summarize my data. These data might give me statistics such as the average number of calories consumed each day by the 100 students in my sample, the most commonly eaten foods, the variety of foods eaten, and so on. All of these statistics simply *describe* characteristics of my sample, and are therefore called **descriptive statistics**. Descriptive statistics generally are used only to describe a specific sample. When all we care about is describing a specific sample, it does not matter whether the scores from the sample are normally distributed or not.

Many times, however, researchers want to do more than simply describe a sample. Sometimes, they want to know what the exact probability is of something occurring in their sample just due to chance. For example, if the average student in my sample consumes 2,000 calories a day, what are the chances, or probability, of having a student in the sample who consumes 5,000 calories a day? In addition, researchers often want to be able to make inferences about the population based on the data they collect from their sample. To determine whether some phenomenon observed in a sample represents an actual phenomenon in the population from which the sample was drawn, **inferential statistics** are used. For example, if I find that the men in my sample eat an average of 200 more calories a day than the women, I may want to know the likelihood, or probability, that this gender difference that I have in my sample represents a real difference between the larger populations of men and women in their first year of college. To calculate these probabilities, I need to rely on the normal distribution, because the characteristics of the normal distribution allow statisticians to generate exact probability statistics. In the next section, I will briefly explain how this works.

THE NORMAL DISTRIBUTION IN DEPTH

It is important to note that the normal distribution is what is known in statistics as a **theoretical distribution**. That is, one rarely, if ever, gets a distribution of scores from a sample that forms an exact, normal distribution. Rather, what you get when you collect data is a distribution of scores that may or may not approach a normal, bell-shaped curve. Because the theoretical normal distribution is what statisticians used to develop the probabilities discussed earlier, a distribution of scores that is not normal may be at odds with these probabilities. Therefore, there are a number of statistics that begin with the assumption that scores are normally distributed. When this assumption is violated (i.e., when the scores in a distribution are not normally distributed), there can be dire consequences.

The most obvious consequence of violating the assumption of a normal distribution is that the probabilities associated with a normal distribution are not valid. For example, if you have a normal distribution of scores on some variable (e.g., IQ test scores of adults in the United States), you can use the probabilities based on the normal distribution to determine exactly what percentage of the scores in the distribution will be 120 or higher on the IQ test (see Chapter 4 for a description of how to do this). But suppose the scores in our distribution do not form a normal distribution. Suppose, for some reason, we have an unusually large number of high scores (e.g., over 120) and an unusually small number of low scores (e.g., below 90) in our distribution. If this were the case, when we use probability estimates based on the normal distribution, we would underestimate the actual number of high scores in our distribution and overestimate the actual number of low scores in our distribution.

Sampling Issues

The problem of violating the assumption of normality becomes most problematic when our sample is not an adequate representation of our population. In most social science research, we do not have access to an entire population. Therefore, we usually need to rely on a sample for our data, and we must hope that our **sample** represents our **population**. Of course, to select a sample that represents our population, we must begin by very carefully defining our population. For example, if I conduct a study in California to examine students' knowledge of world geography, and I select a sample of 500 fifth-grade students from two elementary schools in Los Angeles, what is my population? Do they represent all fifth graders in those two schools? Or all fifth graders in Los Angeles? Or all fifth graders in public schools in Los Angeles? Or in the state? Or in the country?

It turns out that this is a very difficult question to answer. There are all kinds of reasons why a sample may not be a good representation of the larger population that we think it is supposed to represent. Returning to my sample of 500 fifth graders from Los Angeles, suppose that I am trying to make the claim that this sample represents all fifth grade students in Los Angeles public schools. Why might this not be true? For one thing, the two schools from which these students were selected may be different from the typical Los Angeles public elementary school. They may have a larger number of wealthy students than the average school, for example. Or they may spend an unusually small amount of time discussing geography in their fifth grade classrooms. Or they may have an inordinate number of diplomats' children.

On the other hand, these students in my sample may be an excellent representation of the larger population of fifth graders in Los Angeles public schools. Even if they may differ in their wealth or number of students in the classroom or what they eat for breakfast, as long as they do not differ in some *systematic* way *on the variable of interest* (i.e., geography knowledge), there is no reason to believe that this sample does not represent this population.

So the problem of defining our target population is somewhat tricky. Once we have done this, the task becomes one of selecting a sample that has the best chance of representing our population, if possible. One method researchers have adopted is called **random sampling**. Random sampling occurs if members of the sample to be examined in the study are selected *at random* from the population. You should note that the term *at random* has a somewhat different meaning in statistics than in everyday language. It does not mean haphazardly. Rather, it means that every member of the population has an equal chance of being selected for inclusion in the sample. Although this may sound easy, in many cases it is quite difficult. For example, if we define our population as all adults in the United States, how could we make sure that every adult has an equal chance of being selected for our sample? We cannot simply select names out of a phonebook, because not everyone has a phone and those who do not own a phone are more likely to be poor than those who do. We cannot simply select names off of voter registration lists because not everyone is registered to vote, and those who are registered are generally better educated than those who are not. As you can see, it can be difficult to get a truly random sample from a population.

A second method of sampling sometimes used by researchers is called **representative sampling**. In this method, the researcher selects one or more traits to use in selecting a sample from the population. For example, if I am interested in looking at issues of gender and ethnicity, I may try to select a sample that represents the national population (i.e., a sample that is 52% female, 11% African American, etc.). Unfortunately, this can be a costly and time-consuming way of selecting a sample, so many researchers decide against representative sampling.

A third, quite common method of selecting a sample is called **convenience sampling**. In this method, a sample is selected because there was relatively easy access to the members of the sample. For example, if I were to conduct a study of student achievement, I may select a school that is fairly close to where I live or work so that I have less difficulty selecting the data. In addition, human beings have the option of declining participation in a study, so I must select a school that agrees to participate in my study. Once I find a relatively close school that agrees to participate in my study, I may be working with a group of wealthy Lithuanian immigrants who are quite different from the larger population of students in my city, state, and country. With convenience sampling, it is often difficult to define the target population before identifying the sample. Rather, a sample is selected and its characteristics are noted (e.g., race, socioeconomic status, any unique or special characteristics) so that the population that the sample represents can later be inferred.

The Relationship Between Sampling Method and the Normal Distribution

The relationship between the normal distribution and sampling methods is as follows. The probabilities generated from the normal distribution depend on (a) the shape of the distribution, and (b) the idea that the sample is not somehow systematically different from the population. If I select a sample randomly from a population, I know that this sample may not look the same as another sample of equal size selected randomly from the same population. But any differences between my sample and other random samples of the same size selected from the same population would differ from each other randomly, not systematically. In other words, my sampling method was not **biased** such that I would continually select a sample from one end of my population (e.g., the more wealthy, the better educated, the higher achieving) if I continued using the same method for selecting my sample. Contrast this with a convenience sampling method. If I only select schools that are near my home or work, I will continually select schools with similar characteristics. For example, if I live in the Bible Belt, my sample will probably be biased in that my sample will more likely hold fundamentalist religious beliefs than the larger population of schoolchildren. Now if this characteristic is not related to the variable I am studying (e.g., achievement), then it may not matter that my sample is biased in this way. But if this bias is related to my variable of interest, then I may have a problem.

Suppose that I lived and worked in Cambridge, Massachusetts. Cambridge is in a section of the country with an inordinate number of highly educated people because there are a number of high-quality universities in the immediate area (Harvard, MIT, Boston College, Boston University, etc.). If I conducted a study of student achievement using a convenience sample from this area, and tried to argue that my sample represented the larger population of students in the United States, probabilities that are based on the normal distribution may not apply. That is because my sample would be more likely than the national average to score at the high end of the distribution. If, based on my sample, I tried to predict the average achievement level of students in the United States, or the percentage that score in the bottom quartile, or the score that marks the 75th percentile, all of these predictions would be off, because the probabilites that are generated by the normal distribution assume that the sample is not biased. If this assumption is violated, we cannot trust our results.

Skew and Kurtosis

Two characteristics used to describe a distribution of scores are **skew** and **kurtosis**. When a sample of scores is not normally distributed (i.e., the bell shape), there are a variety of shapes it can assume. One way a distribution can deviate from the bell shape is if there is a bunching of scores at one end and a few scores pulling a tail of the distribution out toward the other end. If there are a few scores creating an elongated tail at the higher end of the distribution, it is said to be **positively skewed**. If the tail is pulled out toward the lower end of the distribution, the shape is called **negatively skewed**. These shapes are depicted in Figure 3.2.

As you might have guessed, skewed distributions can distort the accuracy of the probabilities based on the normal distribution. For example, if most of the scores in a distribution occur at the low end with a few scores at the higher end (positively skewed distribution), the probabilities that are based on the normal distribution will underestimate the actual number of scores at the lower end of this skewed distribution and overestimate the number of scores at the higher end of the distribution. In a negatively skewed distribution, the opposite pattern of errors in prediction will occur.

Kurtosis refers the shape of the distribution in terms of height, or flatness. When a distribution has a peak that is higher than that found in a normal, bell-shaped distribution, it is called **leptokurtic**. When a distribution is flatter than a normal distribution, it is called **platykurtic**. Because the normal distribution contains a certain percentage of scores in the middle area (i.e., about 68% of the scores fall between 1 standard deviation above and 1 standard deviation below the mean), a distribution that is either patykurtic or letpkurtic will likely have a different percentage of scores near the mean than will a normal distribution. Specifically, a leptokurtic distribution will probably have a greater percentage of scores closer to the mean and fewer in the upper and lower tails of the distribution whereas a platykurtic distribution will have more scores at the ends and fewer in the middle than will a normal distribution.

Positively Skewed Distribution

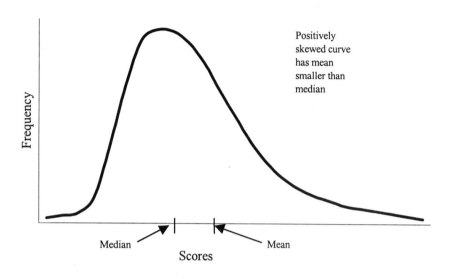

Negatively Skewed Distribution

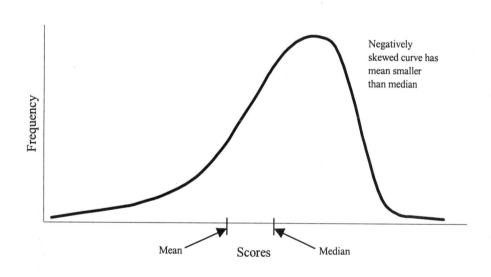

FIGURE 3.2   Positively and negatively skewed distributions.

From *Basic Statistics Tales of Distributions, 6th edition*, by C. Spatz © 1997. Reprinted with permission of Wadsworth, a division of Thomson Learning. Fax 800 730-2215.

EXAMPLE: APPLYING NORMAL DISTRIBUTION PROBABILITIES TO A NON-NORMAL DISTRIBUTION

To illustrate some of the difficulties that can arise when we try to apply the probabilities that are generated from using the normal distribution to a distribution of scores that is skewed, I present a distribution of sixth-grade students' scores on a measure of self-esteem. In these data, 677 students completed a questionnaire that included four items designed to measure students' overall sense of self-esteem. Examples of these questions include "On the whole, I am satisfied with myself" and "I feel I have a number of good qualities." Students responded to each of these four questions using a 5-point rating scale with "1 = not at all true" and "5 = very true." Students' responses on these four items were then averaged, creating a single self-esteem score that ranged from a possible low of 1 to a possible high of 5. The frequency distribution for this self-esteem variable is presented in Figure 3.3.

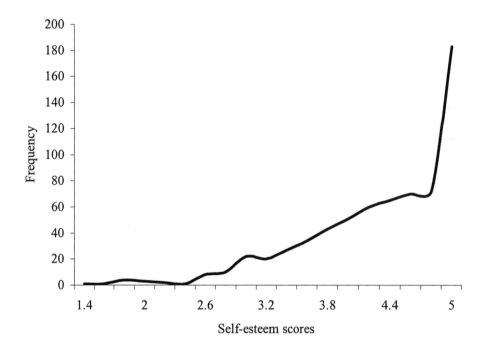

FIGURE 3.3   Frequency distribution for self-esteem scores.

As you can see, the distribution of scores presented in Figure 3.3 does not form a nice, normal, bell-shaped distribution. Rather, most of the students in this sample scored at the high end of the distribution and a long tail extends out toward the lower end of the scale. This is a classic, negatively skewed distribution of scores. The happy part of this story is that most of the students in this sample appear to feel quite good about themselves. The sad part of the story is that some of the assumptions of the normal distribution are violated by this skewed distribution. Let's take a look at some specifics.

One of the qualities of a normal distribution is that it is symmetrical, with an equal percentage of the scores between the mean and 1 standard deviation below the mean as there are between the mean and 1 standard deviation above the mean. In other words, in a normal distribution, there should be about 34% of the scores within 1 standard deviation above the mean and 34% within 1 standard deviation below the mean. In our distribution of self-esteem scores presented earlier, the mean is 4.28 and the standard deviation is .72. A full 50% of the distribution falls between the mean and 1 standard deviation above the mean in this group of scores. So, although I might predict that about 16% of my distribution will have scores more than 1 standard deviation above the mean in a normal distribution, in my skewed distribution of self-esteem scores, I can see that there are no students with scores more than 1 standard deviation above the mean. In Chapter 4,

I present a more thorough discussion of how to use the normal distribution to calculate standard deviation units and percentile scores in a normal distribution.

As this example demonstrates, the probabilities that statisticians have generated using the normal distribution may not apply well to skewed or otherwise non-normal distributions of data. This should not lead you to believe, however, that non-normal distributions of scores are worthless. In fact, even if you have a non-normal distribution of scores in your sample, these scores can create normal sampling distributions for use in inferential statistics (see Chapter 5). What is perhaps most important to keep in mind is that a non-normal distribution of scores may be an indication that your sample may differ in important and systematic ways from the population that it is supposed to represent. When making inferences about a population based on a sample, be very careful to define the population precisely and to be aware of any biases you may have introduced by your method of selecting your sample. The normal distribution can be used to generate probabilities about the likelihood of selecting an individual or another sample with certain characteristics (e.g., distance from the mean) from a population. If your sample is not normal and your method of selecting the sample may be systematically biased to include those with certain characteristics (e.g., higher than average achievers, lower than average income, etc.), then the probabilities of the normal distribution may not apply well to your sample.

WRAPPING UP AND LOOKING FORWARD

The theoretical normal distribution is a critical element of statistics primarily because many of the probabilities that are used in inferential statistics are based on the assumption of normal distributions. As you will see in coming chapters, statisticians use these probabilities to determine the probability of getting certain statistics and to make inferences about the population based on the sample. Even if the data in a sample are not normally distributed, it is possible that the data in the population from which the sample was selected may be normally distributed. In Chapter 4, I describe how the normal distribution, through the use of z scores and standardization, is used to determine the probability of obtaining an individual score from a sample that is a certain distance away from the sample mean. You will also learn about other fun statistics like percentile scores in Chapter 4.

GLOSSARY OF TERMS FOR CHAPTER 3

Asymptotic: When the ends, or "tails" of a distribution never intersect with the X axis. They extend indefinitely.

Bell curve: The common term for the normal distribution. It is called the bell curve because of its bell-like shape.

Biased: When a sample is not selected randomly, it *may* be a biased sample. A sample is biased when the members are selected in a way that systematically overrepresents some segment of the population, and underrepresents other segments.

Convenience sampling: When a sample is selected because it is convenient, rather than random.

Descriptive statistics: Statistics that describe the characteristics of a given sample or population. These statistics are only meant to describe the characteristics of those from whom data were collected.

Kurtosis: The shape of a distribution of scores in terms of its flatness or peakedness.

Leptokurtic: A term regarding the shape of a distribution. A leptokurtic distribution is one with a flatter peak and thicker tails.

Negatively skewed: When a tail of a distribution of scores extends toward the lower end of the distribution.

Normal distribution: A bell-shaped distribution of scores that has the mean, median, and mode in the middle of the distribution and is symmetrical and asymptotic.

Platykurtic: A term regarding the shape of a distribution. A platykurtic distribution is one with a higher peak and thinner tails.

Population: The group from which data are collected or a sample is selected. The population encompasses the entire group for which the data are alleged to apply.

Positively skewed: When a tail of a distribution of scores extends toward the upper end of the distribution.

Probability: The likelihood of an event occurring.

Random sampling: A method of selecting a sample in which every member of the population has an equal chance of being selected.

Representative sampling: A method of selecting a sample in which members are purposely selected to create a sample that represents the population on some characteristic(s) of interest (e.g., when a sample is selected to have the same percentages of various ethnic groups as the larger population).

Sample: An individual or group, selected from a population, from whom or which data are collected.

Skew: The degree to which a distribution of scores deviates from normal in terms of asymmetrical extension of the tails.

Symmetrical: When a distribution has the same shape on either side of the median.

Theoretical distribution: A distribution based on statistical probabilities rather than empirical data.

CHAPTER 4

STANDARDIZATION AND *z* SCORES

If you know the mean and standard deviation of a distribution of scores, you have enough information to develop a picture of the distribution. Sometimes researchers are interested in describing individual scores within a distribution. Using the mean and the standard deviation, researchers are able to generate a **standard score**, also called a *z* **score**, to help them understand where an individual score falls in relation to other scores in the distribution. Through a process of **standardization**, researchers are also better able to compare individual scores in the distributions of two separate variables. Standardization is simply a process of converting each score in a distribution to a *z* score. A *z* score is a number that indicates how far above or below the mean a given score in the distribution is in standard deviation units. So standardization is simply the process of converting individual **raw scores** in the distribution into standard deviation units.

Suppose that you are a college student taking final exams. In your biology class, you take your final exam and get a score of 65 out of a possible 100. In your statistics final, you get a score of 42 out of 200. On which exam did you get a "better" score? The answer to this question may be more complicated than it appears. First, we must determine what we mean by "better." If better means percentage of correct answers on the exam, clearly you did better on the biology exam. But if your statistics exam was much more difficult than your biology exam, is it fair to judge your performance solely on the basis of percentage of correct responses? A more fair alternative may be to see how well you did compared to other students in your classes. To make such a comparison, we need to know the mean and standard deviation of each distribution. With these statistics, we can generate *z* scores.

Suppose the mean on the biology exam was 60 with a standard deviation of 10. That means you scored 5 points above the mean, which is half of a standard deviation above the mean (so higher than the average for the class). Suppose further that the average on the statistics test was 37 with a standard deviation of 5. Again, you scored 5 points above the mean, but this represents a full standard deviation over the average. Using these statistics, on which test would you say you performed better? To fully understand the answer to this question, let's examine standardization and *z* scores in more depth.

STANDARDIZATION AND *Z* SCORES IN DEPTH

As you can see in the previous example, it is often difficult to compare two scores on two variables when the variables were measured using different scales. The biology test was measured on a scale from 1 to 100, whereas the statistics exam used a scale from 1 to 200. When variables have such different scales of measurement, it is almost meaningless to compare the raw scores (i.e., 65 and 42) on these exams. Instead, we need some way to put these two exams on the same scale, or to *standardize* them. One of the most common methods of standardization used in statistics is to convert raw scores into standard deviation units, or *z* scores. The formula for doing this is very simple and is presented in Table 4.1.

As you can see from the formulas in Table 4.1, to standardize a score (i.e., to create a *z* score), you simply subtract the mean from an individual raw score and divide this by the standard deviation. So if the raw score is above the mean, the *z* score will be positive whereas a raw score that is less than the mean will produce a negative *z* score. When an entire distribution of scores is standardized, the average (i.e., mean) z score for the standardized distribution will always be 0, and the standard deviation of this distribution will always be 1.0.

TABLE 4.1 Formula for calculating a z score.

$$z = \frac{raw\ score - mean}{s\tan dard\ deviation}$$

or

$$z = \frac{X - \mu}{\sigma}$$

or

$$z = \frac{X - \overline{X}}{s}$$

where X = raw score

μ = population mean

σ = standard deviation

$\overline{X}$ = sample mean

s = sample standard deviation

Interpreting z scores

z scores tell researchers instantly how large or small an individual score is relative to other scores in the distribution. For example, if I know that one of my students got a z score of –1.5 on an exam, I would know that student scored 1.5 standard deviations below the mean on that exam. If another student had a z score of .29, I would know the student scored .29 standard deviation units above the mean on the exam.

Let's pause here and think for a moment about what z scores do not tell us. If I told you that I had a z score of 1.0 on my last spelling test, what would you think of my performance? What you would know for sure is that (a) I did better than the average person taking the test, (b) my score was 1 standard deviation above the mean, and (c) if the scores in the distribution were normally distributed (Chapter 3), my score was better than about two thirds of the scores in the distribution. But what you would not know would be (a) how many words I spelled correctly, (b) if I am a good speller, (c) how difficult the test was, (d) if the other people taking the test are good spellers, (e) how many other people took the test, and so on. As you can see, a z score alone does not provide as much information as we might want. To further demonstrate this point, suppose that after I told you I had a z score of 1.0 on the spelling test, I went on to tell you that the average score on the test was 12 out of 50 and that everyone else who took the test was 7 years old. Not very impressive in that context, is it?

Now, with the appropriate cautions in mind, let's consider a couple more uses of z scores and standardization. One of the most handy features of z scores is that, when used with a normally distributed set of scores, they can be used to determine **percentile scores**. That is, if you have a normal distribution of scores, you can use z scores to discover which score marks the 90th percentile of a distribution (i.e., that raw score at which 10% of the distribution scored above and 90% scored below). This is because statisticians have demonstrated that in a normal distribution, a precise percentage of scores will fall between the mean and 1 standard deviation above the mean. Because normal distributions are perfectly symmetrical, we know that the exact same percentage of scores that falls between the mean and 1 standard deviation *above* the mean will also fall between the mean and 1 standard deviation *below* the mean. In fact, statisticians have determined the precise percentage of scores that will fall between the mean and any z score (i.e., number of standard deviation units above or below the mean). A table of these values is provided in Appendix A. When you also consider that in a normal distribution the mean always marks the exact center of the distribution, you know that the mean is the spot in the distribution in which 50% of the cases fall below and 50% fall above. With this in mind, it is easy to find the score in a distribution that marks the 90th percentile, or any percentile, for that matter. In Figure 4.1, we can see the percentage of scores in a normal distribution that fall between different z score values. This figure contains the *standard normal distribution*.

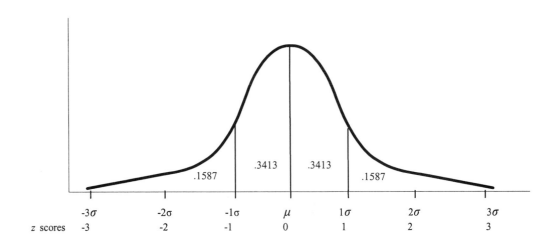

FIGURE 4.1   The standard normal distribution.

Let us consider an example. Suppose I know that the average SAT-math score for White males is 517, with a standard deviation of 100, and forms a normal distribution. In this distribution, I already know that the score that marks the 50th percentile is 517. Suppose I wanted to know the score that marks the 90th percentile. To find this number, I have to follow a series of simple steps.

Step 1: Using a z score table from a traditional statistics textbook, find the z score that marks the 90th percentile. To do this, we need to remember that the 90th percentile is 40 percentile points above the mean, so we are looking for the z score at which 40% of the distribution falls between the mean and this z score. An alternative method is to find the z score at which 10% of the distribution falls above, because the 90th percentile score divides the distribution into sections with 90% of the score falling below this point and 10% falling above. z score tables in traditional statistics textbooks always provide at least one of these scores (i.e., the percentage of the distribution that falls between the mean and the z score or the percentage that falls above the z score), and often both. In the current example, the z score that corresponds with the 90th percentile is 1.28. So $z = 1.28$.

Step 2:   Convert this z score back into the original unit of measurement. Remember that the SAT-math test is measured on a scale from 0 to 800. We now know that the mean for White males who took the test in 1989 was 517, and that 90th percentile score of this distribution is 1.28 standard deviations above the mean (because $z = 1.28$). So what is the actual SAT-math score that marks the 90th percentile? To answer this, we have to convert our z score from standard deviation units into raw score units and add this to the mean. The formula for doing this is

$$X = \mu + (z)(\sigma)$$

In this equation, X is the raw score we are trying to discover, μ is the average score in the distribution, z is the z score we found, and σ is the standard deviation for the distribution. Plugging our numbers into the formula, we find that

$$X = 517 + (1.28)(100)$$
$$X = 517 + 128$$
$$X = 645$$

Step 3: Now we can wrap words around our result and answer our original question. When doing this, it is often helpful to use the original question when stating our finding, as follows:

Question: *What is the score that marks the 90th percentile of the distribution of White male students' SAT-math scores in 1989?*
Answer: *The score of 645 marks the 90th percentile of the distribution of White male students' SAT-math scores in 1989.* This z score, percentile score, and the corresponding raw score are depicted in Figure 4.2.

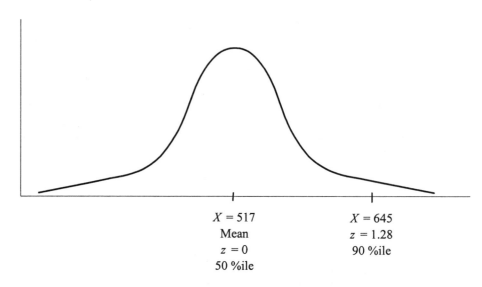

FIGURE 4.2 Finding the score that marks the 90th percentile of the distribution.

Just as we can use z scores to find the raw score that marks a certain percentile in a distribution, we can also use z scores to help us convert a known raw score into a percentile score. For example, if I know that a student in my distribution has a score of 425 on the SAT-math test, I might want to know the percentage of the distribution that scored above and below 425. This is the type of conversion that has happened when students' standardized test scores are published in the local newspaper using percentiles under headlines such as "California Students Score in 45th

Percentile on National Test!" Similarly, when a proud parent exclaims "My Johnny is in the top 10% in height for his age group!" a conversion from a raw score to a percentile score has taken place, with the help of a z score. Here's how it's done:

Step 1: We must begin by converting the raw score into a z score. In our example, the raw score is 425 ($X = 425$). To convert this into a z score, we simply recall our mean ($\mu = 517$) and our standard deviation ($\sigma = 100$) and then plug these numbers into the z score formula:

$$z = \frac{425 - 517}{100}$$

$$z = \frac{-88}{100}$$

$$z = -.88$$

Step 2: Now that we have a z score, we need to look in the nearest z score table to find either the percentage of the normal distribution that falls between the mean and a z score of -.88 or the percentage of the distribution that falls below a z score of -.88. Notice that we are dealing with a negative z score in our example. Most z score tables only report positive z scores, but because normal distributions are symmetrical, the percentage of the distribution that fall between the mean and z are identical whether the z score is positive or negative. Similarly, the percentage that falls above a positive z score is identical to the percentage that falls below a negative z score. My z score table tells me that 31% of the normal distribution of scores will fall between the mean and a z score of -.88 and 19% will fall below a z score of -.88.

Step 3: To wrap words around this result, I must begin with the recollection that in my example, a z score of -.88 corresponds with a raw score of 425 on the SAT-math test among the White males who took the test in 1989. So, to wrap words around my result, I would say that "A score of 425 on the SAT-math test marks the 19th percentile of the distribution of test scores among White males in 1989."

The examples just presented represent two handy uses of z scores for understanding both an entire distribution of scores and individual scores within that distribution. Standardized scores are used in a variety of statistics, and are perhaps most helpful for comparing scores that are measured using different scales of measurement. As discussed earlier in this chapter, it is difficult to compare two scores that are measured on different scales (e.g., height and weight) without first converting them into a common unit of measurement. Standardizing scores is simply this process of conversion. In the final section of this chapter, I present and briefly describe two distributions of scores described by both raw scores and z scores.

EXAMPLES: COMPARING RAW SCORES AND Z SCORES

To illustrate the overlap between raw scores and standardized z scores, I first present data from a sample of elementary and middle school students from whom I collected data a few years ago. I gave these students a survey to assess their motivational beliefs and attitudes about a standardized achievement test they were to take the following week. One of the items on the survey read "The ITBS test will measure how smart I am." Students responded to this question using an 8-point scale with "1 = Strongly Disagree" and "8 = Strongly Agree." The frequency distribution is presented in Figure 4.3. This distribution has a mean of 5.38 and a standard deviation of 2.35.

As you can see, this is not a normal, bell-shaped distribution. This distribution has a sort of odd shape where there is the hint of a normal distribution in Scores 2 through 7 but then there are "spikes" at the ends, particularly at the high end. The result is an asymmetrical distribution. If you compare the z scores on top of each column with the raw scores at the bottom of each column, you can see how these scores are related to each other. For example, we can see that all of the raw scores of 5 or lower have negative z scores. This is because the mean of a distribution always has a z score of 0, and any raw scores below the mean will have negative z scores. In this distribution, the mean is 5.38, so all raw scores of 5 and below have negative z scores and all raw scores of 6 or above have positive z scores.

Another feature of this distribution that is clearly illustrated by the z scores is that there is a larger range of scores below the mean than there are above the mean. This is fairly obvious, because the mean is well above the midpoint on this scale. The highest scores in this distribution are just a little more than 1 standard deviation above the mean ($z = 1.12$), whereas the lowest scores are nearly 2 standard deviations below the mean ($z = -1.86$). Finally, the inclusion of standard deviation scores with each raw score allows us to immediately determine how many standard deviations away from the mean a particular raw score falls. For example, we can see that a student who had a raw score of 3 on this variable scored just about exactly 1 standard deviation below the mean ($z = -1.01$).

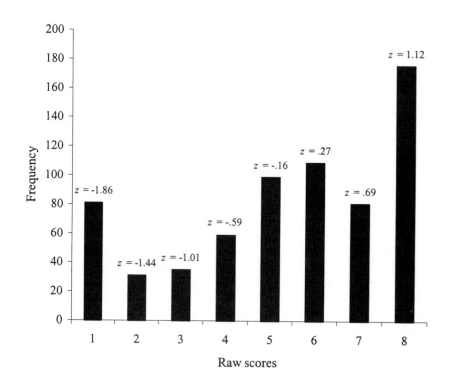

FIGURE 4.3 Frequency distribution for "The test will show how smart I am" item.

For our second example, I have chosen a variable with a much smaller standard deviation. Using the same 8-point scale described earlier, students were asked to respond to the item "I think it is important to do well on the ITBS test." Students overwhelmingly agreed with this statement, as the mean (7.28) and relatively small standard deviation (1.33) revealed. The frequency distribution for the scores on this item is presented in Figure 4.4.

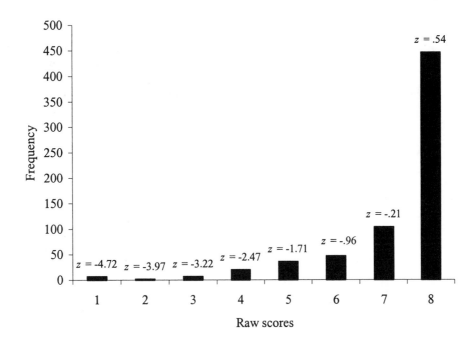

FIGURE 4.4 Frequency distribution for "Important to do well" item.

In this graph, we can see that the distribution is highly skewed, with most students circling the number 8 on the scale. Because so many students answered similarly, the standard deviation is quite small, with only a relatively few scores at the lower end of the distribution. The small standard deviation coupled with the high mean create a situation where very low scores on the scale have extremely small z scores. For example, the few students with a raw score of 1 on the scale (n = 7) had z scores of –4.72, indicating that these students were more than 4   standard deviations below the mean. Those students with the highest score on the scale were only about half a standard deviation above the mean because, with such a high mean, it was impossible to get a score very far above the mean.

The two examples provided previously both illustrate the relation between z scores and raw scores for distributions that are skewed. In both of these distributions, the means were above the midpoint on the scale, and subsequently there was a greater range of z scores below the mean than there were above the mean. Such is not the case when the distribution of scores are normally distributed. To illustrate this, I use data from a different data set. I used surveys to measure a sample of high school students' motivational goals in school. One goal that I measured is known as a performance-approach goal. This goal reflects a concern, or a desire, to outperform classmates and peers for the sake of demonstrating superior ability. The items on the survey were measured using a scale from 1 to 5 (1 = "not at all true" and 5 = "very true"). The frequency distribution is presented in Figure 4.5.

This distribution of scores had a mean of 3.00 and a standard deviation of .92. As you can see, the data are quite normally distributed. When the data are normally distributed, we would expected most of our cases to have z scores at or near zero, because in a normal distribution most of the cases are near the mean. Also notice that as we move farther away from the mean (i.e., z scores over 2.0 or less than –2.0), there are fewer cases. In a normal distribution, then, the probability of finding a particular z score becomes smaller as the value of the z score moves further away from zero. As Figures 4.3 and 4.4 illustrate, this is not always the case in skewed distributions.

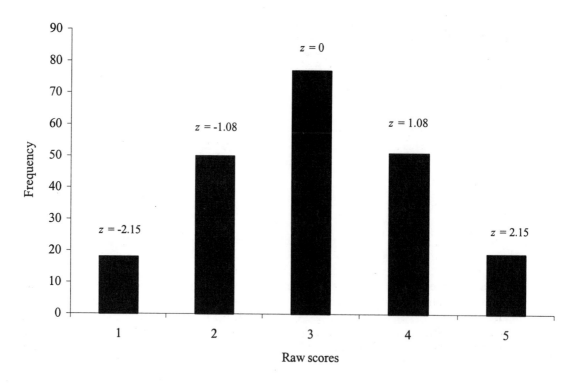

FIGURE 4.5 Frequency distribution for performance-approach goals.

WRAPPING UP AND LOOKING FORWARD

z scores provide a handy way of interpreting where a raw score is in relation to the mean. They also demonstrate that the range of possible *z* scores depends heavily on where the mean falls in the range of possible scores as well as the size of the standard deviation for the distribution. We can use *z* scores to quickly and easily determine where an individual score in a distribution falls relative to other scores in the distribution, either by interpreting the *z* score in standard deviation units or by calculating percentile scores. Using the table of probabilities based on the normal distribution presented in Appendix A, we can also use *z* score to determine how unusual a given score in a distribution is (i.e., the probability of obtaining an individual score of that size when selecting the individual at random). In the next chapter, we will use the information that we have learned about the mean, standard deviation, normal distributions, *z* scores, and probability to understand one of the most important concepts in statistics: the standard error.

GLOSSARY OF TERMS AND SYMBOLS FOR CHAPTER 4

Raw scores: These are the individual observed scores on measured variables.

Standard score: A raw score that has been converted to a *z* score by subtracting it from the mean and dividing by the standard deviation of the distribution. It is an individual score expressed as a deviation from the mean in standard deviation units.

Standardization: The process of converting a raw score into a standard score.

z score: Another term for a standard score.

z The symbol for a standard score.

X The symbol for a raw score.

μ The symbol for a population mean.

σ The symbol for a population standard deviation.

CHAPTER 5

STANDARD ERRORS

The concept of **standard error** is one that many students of statistics find confusing when they first encounter it. In all honesty, there are many students, and many researchers, who never fully grasp the concept. I am convinced that many people have problems with understanding standard errors because they require a bit of a leap into the abstract and because, with the advent of computer programs, it is possible to lead a long and productive research life without having to think about or analyze a standard error for years at a time. Therefore, many researchers choose to gloss over this abstract concept. This is a mistake. I hold this opinion because, as a teacher of statistics, I have learned that when one is able to truly understand the concept of standard error, many of our most beloved inferential statistics (*t* tests, ANOVA, regression coefficients, correlations) become easy to understand. So let me offer this piece of advice: Keep trying to understand the contents of this chapter, and other information you get about standard errors, even if you find it confusing the first or second time you read it. With a little effort and patience, you can understand standard errors and many of the statistics that rely on them.

What Is a Standard Error?

There are two answers to this question. First, there is the technical answer, which is the definition of a standard error. A standard error is, in effect, the standard deviation of the **sampling distribution** of some statistic (e.g., the mean, the difference between two means, the correlation coefficient, etc.). I realize that this makes no sense until you know what a sampling distribution is, and I explain this in the next section of this chapter. For now, I recommend that you repeat the definition to yourself 10 times: "The standard error is, in effect, the standard deviation of the sampling distribution of some statistic." The second answer is that the standard error is the denominator in the formulas used to calculate many inferential statistics. In the following chapters, you will see the standard error as the denominator in many formulas. This is because the standard error is *the measure of how much* random *variation we would expect from samples of equal size drawn from the same population.* Again, look at the preceding sentence, think about it, and rest assured that it is explained in more detail in the next few pages.

STANDARD ERRORS IN DEPTH

The Conceptual Description of the Standard Error of the Mean

To begin this more detailed discussion of standard errors, I introduce the esoteric component of the concept. This is the section that you may need to read several times to let it sink in. Although there are standard errors for all statistics, we will focus on the standard error of the mean.

When we think of a distribution of scores, we think of a certain number of scores that are plotted in some sort of frequency graph to form a distribution (see Chapters 1 and 3). In these distributions, each case has a score that is part of the distribution. Just as these simple frequency distributions are plotted, or graphed, we can also plot distributions of sample means. Imagine that we wanted to find the average shoe size of adult American women. In this study, the population we are interested in is all adult American women. But it would be expensive and tedious to measure the shoe size of all adult American women. So we select a sample of 100 women, at random, from our population. At this point, it is very important to realize that our sample of 100 women may or may not look like the typical American woman (in terms of shoe size). When we select a sample at random, it is possible to get a sample that represents an extreme end of the population (e.g., a sample with an unusually large average shoe size). If we were to throw our first sample of women back into the general population and chose another random sample *of the same size* (i.e., 100), it is possible that this second sample may have an average shoe size that is quite different from our first sample.

Once you realize that different random samples of equal size can produce different mean scores on some variable (e.g., different average shoe sizes), the next step in this conceptual puzzle is easy: If we were to take 1,000 different random samples of women, each of 100, and compute the average show size of each sample, these 1,000 sample means would form their own distribution. This distribution would be called the **sampling distribution of the mean**.

To illustrate this concept, let's consider an example with a small population ($N = 5$). Suppose my population consists of five college students enrolled in a seminar on statistics. Because it is a small seminar, these five students represent the entire population of this seminar. These students each took the final exam that was scored on a scale from 1 to 10, with lower scores indicating poorer performance on the exam. The scores for each student are presented in Table 5.1, arranged in descending order according to how well they did on the exam.

TABLE 5.1   Population of students' scores on final exam.

Students	Score on Final Exam
Student 1	3
Student 2	6
Student 3	6
Student 4	7
Student 5	9

If I were to select a random sample of two students from this population ($n = 2$), I might get student 2 and student 5. This sample would have a mean of 7.5 because [$(6 + 9) \div 2 = 7.5$]. If I were to put those two students back into the population and randomly select another sample of 2, I might get Student 4 and Student 5. This sample would have a mean of 8 because [$(7 + 9) \div 2 = 8$]. I put those students back into the population and randomly select another sample of 2, such as Students 1 an 3. This sample would have a mean of 4.5. As you can see, just by virtue of whom is included in each random sample I select from my population, I get different sample means. Now if I were to repeat this process of randomly selecting samples of two students from my population, calculating their mean, and returning the members of the sample to the population (called **sampling with replacement**), eventually I would get a distribution of sample means that would look something like the distribution presented in Figure 5.1. As you can see, these means form a distribution. This example illustrates how random samples of a given size selected from a population will produce a distribution of sample means, eventually forming a sampling distribution of the mean.

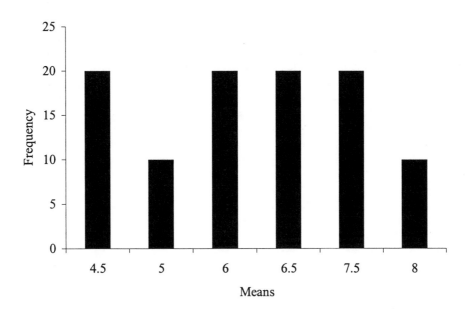

FIGURE 5.1 Sampling distribution of the mean.

Just as the other distributions we have discussed have a mean and a standard deviation, this sampling distribution of the mean also has these characteristics. To distinguish a sampling distribution from a simple frequency distribution, the mean and standard deviation of the sampling distribution of the mean have special names. The mean of the sampling distribution of the mean is called the **expected value**. It is called the expected value because the mean of the sampling distribution of the means is the same as the population mean. When we select a sample from the population, our best guess is that the mean for the sample will be the same as the mean for the population, so our *expected* mean will be the population mean. The standard deviation of the sampling distribution of the mean is called the standard error. So the standard error is simply the standard deviation of the sampling distribution.

The final step in understanding the concept of standard error of the mean is to understand what this statistic tells us. If you will recall the discussion about standard deviations in Chapter 2, you will remember that the standard deviation tells us the average difference, or deviation, between an individual score in the distribution and the mean for the distribution. The standard error of the mean provides essentially the same information, except it refers to the average difference between the expected value (i.e., the population mean) and an individual sample mean. So one way to think about the standard error of the mean is that it tells us how confident we should be that a sample mean represents the actual population mean. Phrased another way, the standard error of the mean provides a measure of how much *error* we can expect when we say that a sample mean represents the mean of the larger population. That is why it is called a standard *error*.

How to Calculate the Standard Error

Most of the time, researchers do not draw 1,000 samples of equal size from the population and then figure out the mean and standard deviation of this distribution of sample means. In fact, most of the time, researchers collect data from only a single sample, and then use this sample to make inferences about the population from which the sample was drawn. How can we make inferences about a larger population on the basis of a single sample?

To make such inferences about the population from a single sample, researchers must use what they know about their sample to make educated guesses, or estimates, about the population. I demonstrate this concept using the shoe-size example mentioned earlier. Suppose that I have a random sample of 100 women. Now if this sample were truly selected at random (i.e., every adult woman in the United States had an equal chance of being selected), my most logical assumption would be that this sample represents the larger population accurately. Therefore, I would have to assume that the mean shoe size of my sample (suppose it is 6) is also the mean shoe size of the larger population. Of course, I cannot know if this is true. In fact, as discussed earlier, there is good reason to believe that my sample may not represent my population well. But if the only information I have about U.S. adult women's shoe size comes from my sample of 100 women, my *best guess* about what the larger population of women looks like must be that they are similar to this sample of 100 women. Now I am faced with a critical question: When I guess that the population of women in the United States has an average shoe size of 6 (based on my sample average), how much *error* can I *expect* to have in this estimation? In other words, what is the *standard error*?

To answer this question I must examine two characteristics of my sample. First, how large is my sample? The larger my sample, the less error I should have in my estimate about the population. This makes sense because the larger my sample, the more my sample should look like my population, and the more accurate my estimates of my population will be. If there are 100 million women in the United States and I use a sample of 50 million to predict their average shoe size, I would expect this prediction to be more accurate than a prediction based on a sample of 100 women. Therefore, the larger my sample, the smaller my standard error.

The second characteristic of my sample that I need to examine is the standard deviation. Remember that the standard deviation is a measure of how much variation there is in the scores in my sample. If the scores in my sample are very diverse (i.e., a lot of variation, a large standard deviation), I can assume that the scores in my population are also quite diverse. So, if some women in my sample have size 2 shoes and others have size 14 shoes, I can assume there is also a pretty large variety of shoe sizes in my population. On the other hand, if all of the women in my sample have shoe sizes of either 5, 6, or 7, I can assume that most of the women in the larger population have an equally small variety of shoe sizes. Of course, these assumptions about the population may not be true, I must rely on them because this is all the information I have. So, the larger the sample standard deviation, the greater the assumed variation of scores in the population, and consequently the larger the standard error of the mean. (**Note:** In those instances where I know the population standard deviation, I can use that in my calculation of the standard error of the mean. See Table 5.2 for that formula.)

TABLE 5.2 Formulas for calculating the standard error of the mean.

$$\sigma_{\bar{x}} = \frac{\sigma}{\sqrt{n}}$$

or

$$s_{\bar{x}} = \frac{s}{\sqrt{n}}$$

where σ is the standard deviation for the population
s is the sample estimate of the standard deviation
n is the size of the sample

An examination of the formula for calculating the standard error of the mean reveals the central role of the sample standard deviation (or population standard deviation, if known) and the sample size in determining the standard error. As you can see, the formula is simply the standard deviation of the sample or population divided by the square root of n, the sample size. As with all fractions, as the numerator gets larger, so does the resulting standard error. Similarly, as the size of the denominator decreases, the resulting standard error increases. Small samples with large standard deviations produce large standard errors, because these characteristics make it more difficult to have confidence that our sample accurately represents our population. In contrast, a large sample with a small standard deviation will produce a small standard error, because such characteristics make it more likely that our sample accurately represents our population.

The Central Limit Theorem

Simply put, the **central limit theorem** states that as long as you have a reasonably large sample size (e.g., $n \geq 30$), the sampling distribution of the mean will be normally distributed, even if the distribution of scores in your sample is not. In earlier chapters (i.e., Chaps. 1 and 3), I discussed distributions that were not in the shape of a nice, normal, bell curve. What the central limit theorem proves is that even when you have such a non-normal distribution in your sample, the sampling distribution of the mean will form a nice, normal, bell-shaped distribution as long as you have at least 30 cases in your sample. Even if you have fewer than 30 cases in your sample, the sampling distribution of the mean will probably be near normal if you have at least 10 cases in your sample. Even in our earlier example where we had only two cases per sample, the sampling distribution of the mean had the beginning of a normal shape.

Although we do not concern ourselves here with why the central limit theorem works, you need to understand why the veracity of this theorem is so important. As I discussed in Chapter 3, a number of statistics rely on probabilities that are generated from normal distributions. For example, I may want to know whether the average IQ test scores of a sample of 50 adults in California is different from the larger population of adults. If my sample has an average IQ test score of 110, and the national average is 100, I can see that my sample average differs from the population average by 10 points. Is 10 points a meaningful difference or a trivial one? To answer that question, I must be able to discover the probability of getting a difference of 10 points by random chance. In other words, if I were to select another random sample of 50 adults from California and compute their average IQ test score, what are the odds that they will have an average that is 10 points higher than the national average of 100? To determine this probability, I must have a normal distribution of sample means, or a normal sampling distribution of the mean. The central limit theorem indicates that as long as I have a sample size of at least 30, my sampling distribution of the mean will be normal.

The Normal Distribution and *t* Distributions: Comparing *z* Scores and *t* Values

In Chapter 4, we learned how to determine the probability of randomly selecting an individual case with a particular score on some variable from a population with a given mean on that variable. We did this by converting the raw score into a *z* score. Now that we know how to compute a standard error, we can use *z* scores again to determine the probability of randomly selecting a sample with a particular mean on a variable from a population with a given mean on the same variable. We can also use the family of *t* distributions to generate *t* values to figure out the same types of probabilities. To explain this, I will begin by comparing the normal distribution with the family of *t* distributions.

As discussed in Chapter 3, the normal distribution is a theoretical distribution with a bell shape and is based on the idea of population data. We also know that the probabilities associated with *z* scores are associated with the normal distribution (Chapter 4). In addition, we know that a standard deviation derived from sample data is only an *estimate* of the population standard deviation (Chapter 2). Because the formula for calculating the sample standard deviation has $n - 1$ in the denominator, we also know that the smaller the sample, the less precisely the sample standard deviation estimates the population standard deviation. Finally, we know that the standard error formula (Table 5.2) is based partly on the standard deviation.

When we put all of this information together, we end up with a little bit of the dilemma. If we can use the standard error to generate *z* scores and probabilities, and these *z* scores and probabilities are based on the normal distribution, what do we do in those cases where we are using sample data and we have a small sample? Won't our small sample influence our standard error? And won't this standard error influence our *z* scores? Will our *z* scores and probabilities be accurate if we have a small sample? Fortunately, these concerns have already be addressed by brains larger than mine. It turns out that the normal distribution has a close family of relatives: the family of *t* distributions. These distributions are very much like the normal distribution, except the shape of *t* distributions is influenced by sample size. With large samples (e.g., > 120), the shape of the *t* distribution is virtually identical to the normal distribution. As sample size decreases, however, the shape of the *t* distribution becomes flatter in the middle and higher on the ends. In other words, as sample size decreases, there will be fewer cases near the mean and more cases away from the mean, out in the tails of the distribution. Like the normal distribution, *t* distributions are still symmetrical.

Just as we use the *z* table (Appendix A) to find probabilities associated with the normal distribution, we use the table of *t* values (Appendix B) to find probabilities associated with the *t* distributions. Along the left column of Appendix B there are numbers in ascending order. These are **degrees of freedom** and they are directly related to sample size. To use this table, you simply calculate a *t* value (using basically the same formula that you use to find a *z* score) and then, using the appropriate degrees of freedom, figure out where your *t* value falls in Appendix B to determine the probability of finding a *t* value of that size. Whenever you don't know the population standard deviation and must use an estimate from a sample, it is wise to use the family of *t* distributions. Here is an example to illustrate these ideas.

In Chapter 4, we used this formula to calculate a *z* score from a raw score:

$$z = \frac{raw\,score - mean}{s\tan dard\,deviation}$$

or

$$z = \frac{X - \mu}{\sigma}$$

where X = raw score
   μ = population mean
   σ = standard deviation

The formula for converting a sample mean into a *z* score is almost identical, except the individual raw score is replaced by the sample mean and the standard deviation is replaced by the standard error. In addition, if we do not know the population standard deviation, the standard deviation estimate from the sample must be used and we are computing a *t* value rather than a *z* score. These formulas are found in Table 5.3.

Now, suppose that I know that the average American man exercises for 60 minutes a week. Suppose, further, that I have a random sample of 144 men and that this sample exercises for an average of 65 minutes per week with a standard deviation of 10 minutes. What is the probability of getting a random sample of this size with a mean of 65 if the actual population mean is 60 *by chance*? To answer this question, I compute a *t* value:

$$t = \frac{65 - 60}{10 \Big/ \sqrt{144}}$$

$$t = \frac{5}{.83}$$

$$t = 6.02$$

If we look in Appendix B, using the row with ∞ degrees of freedom, we can see that the probability of getting a *t* value of this size *by chance* with a sample of this size is less than .001. Notice that if we had calculated a *z* score rather than a *t* score (i.e., if the *population* standard deviation had been 10), our *z* value would have been the same (i.e., *z* = 6.02) and our probability, as found in Appendix A, would have been less than .00003. The normal distribution (associated with *z* scores) and the *t* distributions are virtually identical when the sample size is larger than 120.

Finally, to illustrate the difference between the t distributions and the normal distribution, suppose that our sample size had been 25 rather than 144. We would have calculated the t value just as we did before, but our standard error would be different (because our sample size is smaller), thereby producing a smaller t value:

$$t = \frac{65 - 60}{10 / \sqrt{25}}$$

$$t = \frac{5}{2}$$

$$t = 2.50$$

Now, looking at our table of t values with 24 degrees of freedom, we find that the probability of getting a t value of this size is just about .02. Notice that if we had our larger sample size of 144, the probability of getting a t value of 2.50 would have been closer to .01.

TABLE 5.3 z score and t value formulas.

When σ is known

$$z = \frac{sample\,mean - population\,mean}{standard\,error}$$

or

$$z = \frac{\overline{X} - \mu}{\sigma_{\overline{x}}}$$

When σ is not known

$$z = \frac{sample\,mean - population\,mean}{standard\,error}$$

or

$$z = \frac{\overline{X} - \mu}{s_{\overline{x}}}$$

where μ = population mean

σ^2 = standard error

$\overline{X}$ = sample mean

$s_{\overline{x}}$ = sample estimate of the standard error

So when the sample size is large, the normal distribution and the t distribution are virtually identical. But as our sample size decreases, the t distribution changes and so do the probabilities associated with it. When the population standard deviation is known, the normal distribution can be used. But when the population standard deviation is not know, or the sample size is small, the family of t distributions should be used. Now we can turn our attention the how standard errors are used in other statistics.

The Use of Standard Errors in Inferential Statistics

Some type of standard error is used in every inferential statistic, including those discussed in this book (*t* tests, ANOVA, regression, etc.). In **inferential statistics**, we want to know whether something that we observe in our sample represents a similar phenomenon in the larger population from which the sample was drawn. For example, if I compare the average weight of a sample 100 men to that of a sample of 100 women and find that, on average, men in my sample weigh 60 pounds more than women in my sample, I may want to know whether I should conclude that, on average, men in the larger population weigh more than women in the population. Similarly, if I find a correlation (see Chapter 7) of *r* = .45 between height and weight in my sample of 100 men, I might want to know whether this relationship between height and weight in my sample means there is probably a relationship between these two variables in the larger population of men. To answer these questions, I need to use standard errors.

In many inferential statistics formulas, I need to see whether the phenomenon I observed in my sample(s) is large or small relative to my standard error. Recall from the definition of standard error presented earlier in this chapter that a standard error is a measure of the average amount of variance, or difference, we can expect from different samples of the same size selected from a population. So, the question we are asking with many inferential statistics is whether some statistic we see in our sample is big or small compared to the amount of variance (or error) we would expect if we had randomly selected a *different* sample of the same size. This question can be summarized with the following fraction:

$$\frac{size\ of\ sample\ statistic}{s\tan dard\ error}$$

As an illustration, let us return to the example comparing the average weight of men and women. We already know that, in my samples, the difference between the average weight of men and women was 60 pounds. The statistic that I am interested in here is the *difference* between the two means (i.e., the average weight of men and the average weight of women). If I were to select two different samples of the same size from the populations of men and women and find the difference in those two sample means, I would probably find a difference that was either larger or smaller than the difference I found in the comparison of the first two samples. If I kept selecting different samples and compared their means, I would eventually get a **sampling distribution of the differences between the means**, and this sampling distribution would have a standard error. Suppose that the standard error of this sampling distribution was 10. Let's plug that standard error into our fraction formula presented earlier:

$$\frac{sample\ statistic = 60}{s\tan dard\ error = 10}$$

From this formula I can see that the difference between my two sample means is six times larger than the difference I would expect to find just due to random sampling error. This suggests that the difference between my two sample means is probably not due to chance. Using a table of probabilities based on the *t* distribution (see Chapter 8 and Appendix B), I can calculate the exact probability of getting a ratio this large (i.e., 60:10, or 6:1). So, to summarize, the standard error is often used in inferential statistics to see whether our sample statistic is larger or smaller than the average differences in the statistic we would expect to occur by chance due to differences between samples. I now discuss some examples to demonstrate the effect of sample size and the standard deviation on the size of the standard error of the mean.

EXAMPLE: SAMPLE SIZE AND STANDARD DEVIATION EFFECTS ON THE STANDARD ERROR

To illustrate the effect that sample size and standard deviation have on the size of the standard error of the mean, let's take a look at a variable from a set of data I collected a few years ago. The purpose of the study was to examine students' motivational beliefs about standardized achievement tests. I examined whether students thought it was important to do well on the standardized test they were about to take in school, whether they had anxiety about the test, whether they expected to do well on the test, whether they thought of themselves as good test takers, and so on.

One of the goals of the study was to compare the motivational beliefs of elementary school students with those of middle school students. The sample for the study included 137 fifth graders in elementary school with 536 seventh and eighth graders in middle school. Suppose we wanted to know the standard error of the mean on the variable "I expect to do well on the test" for each of the two groups in the study, the elementary school students and the middle school students. To calculate these standard errors, we would need to know the standard deviation for each group on our variable and the sample size for each group. These statistics are presented in Table 5.4.

TABLE 5.4 Standard deviations and sample sizes.

	Elementary School Sample		Middle School Sample	
	Standard Dev.	Sample Size	Standard Dev.	Sample Size
Expect to do well on test	1.38	137	1.46	536

A quick glance at the standard deviations for each group reveals that they are very similar (s = 1.38 for the elementary school sample, s = 1.46 for the middle school sample). However, because there is quite a large difference in the size of the two samples, we should expect somewhat different standard errors of the mean for each group. Which group do you think will have the larger standard error of the mean?

Recall from the formula presented earlier in this chapter that to find the standard error of the mean, we simply need to divide the standard deviation by the square root of the sample size. For the elementary school sample, we need to divide 1.38 by the square root of 137. The square root of 137 = 11.70. When we divide 1.38 by 11.70 we get .12. So the standard error of the mean for the elementary sample is .12. Following the same procedure for the middle school sample, we find that the standard error of the mean for this group will equal 1.46 divided by the square root of 546. The square root of 546 = 23.37. When we divide 1.46 by 23.37 we get .06. As you can see, the standard error of the mean for the middle school sample (s.e. = .06) is half the size of the standard error of the mean for the elementary school sample (s.e. = .12). Because the standard deviations were roughly equal for these two groups, virtually all of the difference in their standard errors is attributable to differences in sample size.

To illustrate the effect of the standard deviation on the size of the standard error, let's take a look at a second variable from this study, students' scores on the verbal portion of the standardized achievement tests. Scores on this portion of the test range from a possible low of 0 to a possible high of 100. In the elementary school sample, the standard deviation on this variable was 23.81. The sample size is still 137. To find the standard error of the mean, we must divide 23.81 by the square root of 137 (which we know from our previous example is 11.70. And, 23.81 divided by 11.70 equals 2.04. So the standard error of the mean in this example is 2.04. When we compare this number with the standard error of the mean for the elementary school sample on the "Expect to do well on the test" variable (s_x = .12), we see that the larger standard deviation for the test score variable created a much larger standard error, even though the sample size remained the same, 137.

As these examples demonstrate, the size of the standard error of the mean depends on the size of the standard deviation and the size of the sample. As sample size increases, the standard error of the mean decreases. As the size of the standard deviation increases, the size of the standard error of the mean increases as well. Remember that the standard error is generally used in the denominator of the formulas statisticians use to calculate inferential statistics. Therefore, smaller standard errors will produce larger statistics (because smaller denominators produce larger overall numbers than larger denominators do when the numerators are equal). Larger statistics are more likely to be statistically significant. Therefore, all else being equal, larger sample sizes are more likely to produce statistically significant results because larger sample sizes produce smaller standard errors.

WRAPPING UP AND LOOKING FORWARD

Standard error is often a difficult concept to grasp the first time it is encountered (or the second or the third). Because it is such a fundamental concept in inferential statistics, however, I encourage you to keep trying to make sense of both the meaning and the usefulness of standard errors. As we learned in this chapter, standard errors can be used to determine probabilities of sample statistics (such as the mean) in much the same way that we used standard scores to determine probabilities associated with individual scores in Chapter 4. Because of the usefulness of standard errors in determining probabilities, standard errors play a critical role in determining whether a statistic is statistically significant. Because standard errors are influenced by sample size, statistical significance will also be influenced by these sample characteristics. In the next chapter, the issue of statistical significance, and the effects of sample size on statistical significance, are discussed in more depth.

GLOSSARY OF TERMS AND SYMBOLS FOR CHAPTER 5

Central limit theorem: The fact that as sample size increases, the sampling distribution of the mean becomes increasingly normal, regardless of the shape of the distribution of the sample.

Degrees of freedom: Roughly, the minimum amount of data needed to calculate a statistic. More practically, it is a number, or numbers, used to approximate the number of observations in the data set for the purpose of determining statistical significance.

Expected value: The value of the mean one would expect to get from a random sample selected from a population with a known mean. For example, if one knows the population has a mean of 5 on some variable, one would expect a random sample selected from the population to also have a mean of 5.

Inferential statistics: Statistics generated from sample data that are used to make inferences about the characteristics of the population the sample is alleged to represent.

Sampling distribution of the differences between the means: The distribution of scores that would be generated if one were to repeatedly draw two random samples of a given size from two populations and calculate the difference between the sample means.

Sampling distribution of the mean: The distribution of scores that would be generated if one were to repeatedly draw random samples of a given size from a population and calculate the mean for each sample drawn.

Sampling distribution: A theoretical distribution of any statistic that one would get by repeatedly drawing random samples of a given size from the population and calculating the statistic of interest for each sample.

Standard error: The standard deviation of the sampling distribution.

$s_{\bar{x}}$ Symbol for the standard error of the mean estimated from the sample standard deviation (i.e., when the population standard deviation is unknown).

$\sigma_{\bar{x}}$ Symbol for the standard error of the mean when the population standard deviation is known.

CHAPTER 6

STATISTICAL SIGNIFICANCE
AND EFFECT SIZE

Statistical significance and **effect size** are two of the most fundamental concepts in quantitative research. Both of these provide indexes of how meaningful the results of statistical analyses are. Despite their frequent appearance in reports of quantitative research (particularly measures of statistical significance), too many researchers lack a clear understanding of what precisely what these concepts mean. The purpose of this chapter is to provide a solid foundation of understanding about statistical significance and effect size for you, the reader, to understand these concepts when you encounter them. Because statistical significance and effect size can be calculated for virtually any statistic, it is not possible in this short chapter to provide instructions on how to determine statistical significance or calculate an effect size across all research situations. Therefore, the focus of this chapter is to describe what these concepts mean and how to interpret them, as well as to provide general information about how statistical significance and effect sizes are determined.

Statistics are often divided into two types: **descriptive statistics** and **inferential statistics**. As I mentioned in Chapter 3, descriptive statistics are those statistics that *describe* the characteristics of a given set of data. For example, if I collect weight data for a group of 30 adults, I can use a variety of statistics to describe the weight characteristics of these 30 adults (e.g., their average, or mean, weight, the range from the lowest to the highest weight, the standard deviation for this group, etc.). Notice that all of these descriptive statistics do nothing more than provide information about this specific group of 30 individuals from whom I collected data.

Although descriptive statistics are useful and important, researchers are often interested in extending their results beyond the specific group of people from whom they have collected data (i.e., their sample, or samples). From their sample data, researchers often want to determine whether there is some phenomenon of interest occurring in the larger population(s) that these samples represent. For example, I may want to know whether, in the general population, boys and girls differ in their levels of physical aggression. To determine this, I could conduct a study in which I measure the physical aggression levels of every boy and girl in America and see whether boys and girls differ. This study would be very costly, however, and very time consuming. Another approach is to select a sample of boys and a sample of girls, measure their levels of physical aggression, see if they differ, and from these sample data *infer* about differences in the larger populations of boys and girls. If I eventually conclude that my results are statistically significant, in essence I am concluding that the differences I observed in the average levels of aggression of the boys and girls in my two samples represents a likelihood that there is also a difference in the average levels of aggression in the populations of boys and girls from which these samples were selected.

As the name implies, *inferential* statistics are always about making inferences about the larger population(s) on the basis of data collected from a sample or samples. To understand how this works, we first need to understand the distinction between a population and a sample and get comfortable with some concepts from probability. Once we have developed an understanding of statistical significance, we can then compare the concepts of **statistical significance** and **practical**

significance. This distinction leads us to the second major concept covered in this chapter, which is effect size. Briefly, effect size is a measure of how large an observed effect is without regard to the size of the sample. In the earlier example examining levels of aggression, the effect that I am interested in is the difference in boys' and girls' average levels of aggression.

STATISTICAL SIGNIFICANCE IN DEPTH

Samples and Populations

The first step in understanding statistical significance is to understand the difference between a **sample** and a **population**. This difference has been discussed earlier, both in brief (the Author's Aside in Chapter 1) and in more detail (Chapter 3). Briefly, a sample is an individual or group from whom or from which data are collected. A population is the individual or group that the sample is supposed to represent. For the purposes of understanding the concept of statistical significance, it is critical that you remember that when researchers collect data from a sample, they are often interested in using these data to make inferences about the population from which the sample was drawn. Statistical significance refers to the likelihood, or probability, that a statistic derived from a *sample* represents some genuine phenomenon in the *population* from which the sample was selected. In other words, statistical significance provides a measure to help us decide whether what we observe in our sample is also going on in the population that the sample is supposed to represent.

One factor that often complicates this process of making inferences from the sample to the population is that in many, if not most, research studies in the social sciences, the population is never explicitly defined. This is somewhat problematic, because when we argue that a statistical result is statistically significant, we are essentially arguing that the result we found in our sample is representative of some effect in the population from which the sample was selected. If we have not adequately defined our population, it is not entirely clear what to make of such a result. For the purposes of this chapter, however, suffice it to say that samples are those individuals or groups from whom or which data are collected, whereas populations are the entire collection of individuals or cases from which the samples are selected.

Probability

As discussed earlier in Chapter 3 and Chapter 5, probability plays a key role in inferential statistics. When it comes to deciding whether a result in a study is statistically significant, we must rely on probability to make the determination. Here is how it works.

When we calculate an inferential statistic, that statistic is part of a sampling distribution. From our discussion of standard errors in Chapter 5, you will recall that whenever we select a sample from a population and calculate a statistic from the sample, we have to keep in mind that if we had selected a different sample *of the same size from the same population*, we probably would get a slightly different statistic from the new sample. For example, if I randomly selected a sample of 1,000 men from the population of men in the United States and measured their shoe size, I might find an average shoe size of 10 for this sample. Now, if I were to randomly select a new sample of 1,000 men from the population of men in the United States and calculate their average shoe size, I might get a different mean, such as 9. If I were to select an infinite number of random samples of 1,000 and calculate the average shoe sizes of each of these samples, I would end up with a sampling distribution of the mean, and this sampling distribution would have a standard deviation, called the standard error of the mean (see Chapter 5 for a review of this concept if needed). Just as there is a sampling distribution and a standard error of the mean, so there are sampling distributions and standard errors for all statistics, including correlation coefficients, F ratios from ANOVA, t values from t tests, regression coefficients, and so on.

Because these sampling distributions have certain stable mathematical characteristics, we can use the standard errors to calculate the exact probability of obtaining a specific sample statistic, from a sample of a given size, using a specific known or hypothesized population parameter. It's time for an example. Suppose that, from previous research by the shoe industry, I know that the average shoe size for the population of men in the U.S. is a size 9. Because this is the known average for the *population*, this average is a *parameter* and not a statistic. Now suppose I randomly select a sample of 1,000 men and find that their average shoe size is 10, with a standard deviation of 2. Notice that my average for my sample (10) is a *statistic* because it comes from my sample, not my population. With these numbers, I can answer two slightly different but related questions. First, if the average shoe size in the population is really 9, what is the probability of selecting a random sample of 1000 men who have an average shoe size of 10? Second, is the difference between my population mean (9) and my sample mean (10) *statistically significant*? The answer to my first question provides the basis for the answer to my second question.

Notice that if I do not select my sample at random, it would be easy to find a sample of 1,000 men with an average shoe size of 10. I could buy customer lists from shoe stores and select 1,000 men who bought size 10 shoes. Or I could place an advertisement in the paper seeking men who wear a size 10 shoe. But if my population mean is really 9, and my sample is really selected at random, then there is some probability, or *chance*, that I could wind up with a sample of 1,000 men with an average shoe size of 10. In statistics, this *chance* is referred is as **random sampling error or random chance.**

Back to the example. If my population mean is 9, and my random sample of 1000 men has a mean of 10 and a standard deviation of 2, I can calculate the standard error by dividing the standard deviation by the square root of the sample size (see Chapter 5 for this formula).

$$s_{\bar{x}} = 2 \div \sqrt{1000}$$

$$s_{\bar{x}} = 2 \div 31.62$$

$$s_{\bar{x}} = .06$$

Where $s_{\bar{x}}$ = the standard error of the mean

Now that I know the standard error is .06, I can calculate a t value to find the approximate probability of getting a sample mean of 10 *by random chance* if the population mean is really 9. (**Note**: For sample sizes larger than 120, the t distribution is identical to the normal distribution. Therefore, for large sample sizes, t values and z values, and their associated probabilities, are virtually identical. See Chapters 3 and 5 for more information.)

$$t = \frac{9 - 10}{.06}$$

$$t = \frac{-1}{.06}$$

$$t = -16.67$$

When using the t distribution to find probabilities, we can simply take the absolute value of t. Once we have our absolute value for t ($t = 16.67$), we can consult the t table in Appendix B and see that, when the degrees of freedom equals infinity (i.e., greater than 120), the probability of getting a t value of 16.67 is less than .001. In fact, because the critical t value associated with a probability of .001 is only 3.291, and our actual t value is 16.67, we can conclude that the random chance of getting a sample mean of 10 when the population mean is 9 is *much* less than .001. In other words, when we randomly select a sample of 1,000 men and calculate their average shoe size, when we know that the average shoe size of men in the population is 9, we would *expect* to get a sample mean of 10 much less than 1 time in 1,000. With our table of t values, that is as accurate as we can get.

So we have already calculated the probability, or random chance, of finding a sample mean of 10 when the population mean is 9 is very small, less than one in a thousand, when the sample size is 1,000 and is randomly selected. This probability is known as a ***p* value**, with the ***p*** standing for *probability*. In our current example, we would say that we found a $p < .001$, which is the way *p* values are generally reported in research reports and scholarly journals. Now we can turn our attention to the second question: Is the difference between a population mean of 9 and a sample mean of 10 statistically significant? Well, the quick answer is "Yes." The longer answer requires us to delve into the world of hypothesis testing.

Interpreting SPSS Output

Fortunately, researchers generally do not have to use t distribution tables to calculate probabilities anymore because statistical software programs often provide more precise probability values. In the following, you can find some output from SPSS, a population statistics software program. In this example, I have a sample of 475 high school students. Suppose that I know the average high school students' grade point average (GPA)in the United States is 2.75 on a scale from 0 to 4 (0 = "F" and 4 = "A"). This is about a C+. The average GPA for students in my high school sample is 2.8607. I want to determine the probability of getting a sample mean of 2.8607 *by random chance* if my population mean is really 2.75. I calculate my t statistic and get the following results:

Variable	Number of Cases	Mean	SD	SE of Mean
GPA	475	2.8607	.789	.036

Test Value = 2.75

Mean Difference	95% CI Lower	Upper		t Value	df	2-Tail Sig
.11	.040	.182		3.06	474	.002

The top lines of this output indicate that for our variable called "GPA" we have 475 cases in our sample, a mean of 2.8607, and a standard deviation of .789. Notice that the standard error of the mean ("SE of Mean") is .036, which is simply the standard deviation divided by the square root of the sample size. Next, we have a "Test Value" of 2.75, indicating that we are testing whether our sample mean of 2.8607 is statistically significantly different from the population mean of 2.75. In the bottom lines of the output, we are given the difference between the sample mean and the population mean (2.86 - 2.75 = .11). Next, SPSS provides the confidence interval for this difference between the means. Then we are given a t value of 3.06, followed by degrees of freedom ("*df*"). Finally, we are given a p value of .002 ("2-Tail Sig"). This p value tells us that the probability of obtaining a sample mean of 2.86 from a sample of 475 drawn randomly from the population *by chance* when the actual population mean is 2.75 is two in a thousand (i.e., $p = .002$). Because this is a smaller probability than .05, we would conclude that our sample mean is statistically significantly different from the population mean.

Hypothesis Testing and Type I Errors

In the real world of social science research, the model of testing a p value against some set of null and alternative hypotheses is rarely followed. In fact, in most social science research, the researcher collects data from a sample, runs some statistic of interest, and decides that if the probability of finding the statistic by chance (i.e., the p value) is less than .05 (i.e., $p < .05$), the result is statistically significant. But the decision to proclaim any statistic that has a p value less than .05 "statistically significant" is grounded in a tradition of hypothesis testing that is worth reviewing briefly. Note that extended discussions of hypothesis testing can be found in any standard statistics textbook.

The idea here is simple. Before we decide whether a result is statistically significant, we should establish a standard, or benchmark, *before* we calculate the statistic. To do this, we develop a hypothesis and establish a criterion that we will use when deciding whether to retain or reject our hypothesis. The primary hypothesis of interest in social science research is the **null hypothesis** **(H_o).** As the name implies, the null hypothesis always suggests that there will be an *absence* of effect. To illustrate, let us return to the shoe-size example. Recall that we already knew our population average shoe size was 9. Given this, we would expect that if we were to randomly select a sample from that population, and calculate the average shoe size for the sample, that average would also be 9. We might know that there is a *chance* our sample would have a different mean than our population, but our best guess is that our sample would have the same mean as our population. Therefore, our null hypothesis would be that our population mean and our sample mean would not differ from each other (i.e., no effect). We could write this hypothesis symbolically as follows:

$$H_o: \mu = \overline{X} = 9$$

where μ represents the population mean
$\overline{X}$ represents the sample mean

Notice that at this point, we have not yet selected our sample of 1,000 men and we have not yet calculated a sample mean. This entire hypothesis building process occurs *a priori*. Of course, where there is one hypothesis (the null), it is always possible to have alternative hypotheses. One alternative to the null hypothesis is the opposite hypothesis. Whereas the null hypothesis is that the sample and population means will equal each other, an alternative hypothesis could be that they will not equal each other. This **alternative hypothesis (H_A or H_1)** would be written symbolically as

$$H_A: \mu \neq \overline{X}$$

$$\overline{X} \neq 9$$

where μ represents the population mean
$\overline{X}$ represents the sample mean

At this point in the process, we have established our null and alternative hypotheses. You may assume that all we need to do is randomly select our 1000 men, find their average shoe size, and see if it is different from or equal to 9. But, alas, it is not quite that simple. Suppose that we get our sample and find their average shoe size is 9.00001. Technically, that is different from 9, but is it different enough to be considered meaningful? Keep in mind that whenever we select a sample at random from a population, there is always a chance that it will differ slightly from the population. Although our best guess is that our sample mean will be the same as our population mean, we have to remember that it would be almost impossible for our sample to look *exactly* like our population. So our question becomes this: How different does our sample mean have to be from our population mean before we consider the difference meaningful, or *significant.* If our sample mean is just a little different from our population mean, we can shrug it off and say "Well, the difference is probably just due to *random sampling error*, or *chance.*" But how different do our sample and population means need to be before we conclude that the difference is probably *not* due to chance? That's where our **alpha level,** or **Type I error**, comes into play.

As I explained earlier in this chapter, and in Chapters 3 and 5, sampling distributions and standard errors of these distributions allow us to compute probabilities for obtaining sample statistics of various sizes. When I say "probability" this is, in fact, shorthand for "the probability of obtaining this sample statistic due to *chance* or *random sampling error.*" Given that samples generally do not precisely represent the populations from which they are drawn, we should expect some difference between the sample statistic and the population parameter simply due to the luck of the draw, or random sampling error. Had we reached into our population and pulled out another random sample, we probably would get slightly different statistics again. So some of the difference between a sample statistic, like the mean, and a population parameter will always be due to who we happened to get in our random sample, which is why it is called random sampling error. Recall from Chapter 5 that, with a statistic like the mean, the sampling distribution of the mean is a normal distribution. So our random sampling method will produce many sample means that are close to the value of the population mean and fewer that are further away from the population mean. The further the sample mean is from the population mean, the less likely it is to occur by chance, or random sampling error.

Before we can conclude that the differences between the sample statistic and the population parameter are probably *not* just due to random sampling error, we have to decide how unlikely the chances are of getting a difference between the statistic and the population parameter just by chance. In other words, before we can *reject the null hypothesis*, we want to be reasonably sure that any difference between the sample statistic and the population parameter is not just due to random sampling error, or chance. In the social sciences, the convention has generally been to set that level at .05. In other words, social scientists have generally agreed that if the probability of getting a difference between the sample statistic and the population parameter is less than 5%, we will reject the null hypothesis and conclude that the differences between the statistic and the parameter are probably not due to chance.

The agreed-upon probability of .05 (symbolized as $\alpha = .05$) represents the Type I error rate that we, as researchers, are willing to accept *before we conduct our statistical analysis*. Remember that the purpose of our analysis is to determine whether we should retain or reject our null hypothesis. When we decide to reject the null hypothesis, what we are saying in essence, is that we are concluding that the difference between our sample statistic and our population parameter is *not* due to random sampling error. But when we make this decision, we have to remember that it is always possible to get even very large differences just due to random sampling error, or chance. In our shoe-size example, when I randomly select 1,000 men, it is possible that, just due to some fluke, I select 1,000 men with an average shoe size of 17. Now this is extremely unlikely, but it is always possible. You never know what you're going to get when you select a random sample. In my earlier example, where my sample had an average shoe size of 10, I found the probability of getting a sample mean of 10 when my population mean was 9, by chance, was less than one in a thousand. Though unlikely, *it is still possible* that this difference between my sample and population means were just due to chance. So because my *p* value ($p < .001$) is much smaller than my alpha level ($\alpha = .05$), I will reject the null hypothesis and conclude that my sample mean is actually different from my population mean, that this is probably not just a fluke of random sampling, and that my result is statistically significant. When I reach this conclusion, I may be wrong. In fact, I may be rejecting the null hypothesis, even though the null hypothesis is true. Such errors (rejecting the null hypothesis when it is true) are called Type I errors.

To summarize, when we do inferential statistics, we want to know whether something that we observe in a sample represents an actual phenomenon in the population. So we set up a null hypothesis that there is no real difference between our sample statistic and our population parameter, and we select an alpha level that serves as our benchmark for helping us decide whether to reject or retain our null hypothesis. If our *p* value (which we get *after* we calculate our statistic) is smaller than our selected alpha level, we will reject the null hypothesis. When we reject the null hypothesis, we are concluding that the difference between the sample statistic and the population parameter is probably not due to chance, or random sampling error. However, when we reach this conclusion, there is always a chance that we will be wrong, having made a Type I error. One goal of statistics is to avoid making such errors, so to be extra safe we may want to select a more conservative alpha level, such as .01, and say that unless our *p* value is even smaller than .01, we will retain our null hypothesis. In our shoe-size example, our *p* value was much smaller than either .05 or .01, so we reject the null hypothesis and conclude that, for some reason, our sample of 1,000 men had a statistically significantly larger average shoe size than did our general population.

EFFECT SIZE IN DEPTH

As an indication of the importance of a result in quantitative research, statistical significance has enjoyed a rather privileged position for decades. Social scientists have long given the "$p < .05$" rule a sort of magical quality, with any result carrying a probability greater than .05 being quickly discarded into the trash heap of "non significant" results. Recently, however, researchers and journal editors have begun to view statistical significance in a slightly less flattering light, recognizing one of its major shortcomings: It is perhaps too heavily influenced by sample size. As a result, more and more researchers are becoming aware of the importance of effect size, and increasingly are including reports of effect size in their work.

To determine whether a statistic is statistically significant, we follow the same general sequence regardless of the statistic (z scores, t values, F values, correlation coefficients, etc.). First, we find the difference between a sample statistic and a population parameter (either the actual parameter or, if this is not known, an hypothesized value for the parameter). Next, we divide that difference by the standard error. Finally, we determine the probability of getting a ratio of that size due to chance, or random sampling error. (For a review of this process, refer to the earlier section in this chapter when we calculated the t value for the shoe size example).

The problem with this process is that when we divide the numerator (i.e., the difference between the sample statistic and the population parameter) by the denominator (i.e., the standard error), the sample size plays a large role. In all of the formulas that we use for standard error, the larger the sample size, the smaller the standard error (see Chapter 5). When we plug the standard error into the formula for determining t values, f-values, and z scores, we see that the smaller the standard error, the larger these values become, and the more likely that they will be considered statistically significant. Because of this effect of sample size, we sometimes find that even very small differences between the sample statistic and the population parameter can be statistically significant if the sample size is large.

To illustrate this point, let us consider an example with two different sample sizes. Suppose we know that the average IQ score for the population of adults in the United States is 100. Now suppose that I randomly select two samples of adults. One of my samples contains 25 adults, the other 1600. Each of these two samples produces an average IQ score of 105 and a standard deviation of 15. Is the difference between 105 and 100 statistically significant? To answer this question, I need to calculate a t value for each sample. The standard error for our sample with 25 adults will be

$$\sigma_{\bar{x}} = 15 \div \sqrt{25} \Rightarrow 15 \div 5 \Rightarrow 3$$

where $\sigma_{\bar{x}}$ is the standard error of the mean

The standard error for our second sample, with 1,600 adults, will be

$$\sigma_{\bar{x}} = 15 \div \sqrt{1600} \Rightarrow 15 \div 40 \Rightarrow .375$$

Plugging these standard errors into our t value formulas, we find that the t value for the 25-person sample is 105 – 100/3, or 1.67. Looking in our table of t distributions (Appendix B) we can see that the p value for a t value of 1.67 is between .10 and .20. The t value for the sample with 1,600 adults is 105 – 100/.375, or 13.33, with a corresponding p value of $p < .0001$. If we are using an alpha level of .05, then a difference of 5 points on the IQ test would not be considered statistically significant if we only had a sample size of 25, but would be highly statistically significant if our sample size were 1600. Because sample size plays such a big role in determining statistical significance, many statistics textbooks make a distinction between statistical significance and **practical significance**. With a sample size of 1600, a difference of even 1 point on the IQ test would produce a statistically significant result ($t = 1 \div .375 \Rightarrow t = 2.67, p < 01$). However, if we had a very small sample size of 4, even a 15-point difference in average IQ scores would not be statistically significant ($t = 15 \div 7.50 \Rightarrow t = 2.00, p > 10$). (See Figure 6.1 for a graphic illustration of this.) But is a difference of one point on a test with a range of over 150 points really important in the real world? And is a difference of 15 points not meaningful? In other words, is it a significant difference in the *practical* sense of the word *significant*? One way to answer this question is to examine the effect size.

There are different formulas for calculating the effect sizes of different statistics, but each of these formulas share common features. The formulas for calculating most inferential statistics involve a ratio of a numerator (such as the difference between a sample mean and a population mean in a one-sample t test) divided by a standard error. Similarly, most effect size formulas use the same numerator, but divide this numerator by a standard *deviation* rather than a standard error. The trick, then, is knowing how to come up with the appropriate standard deviation to use in a particular effect size formula.

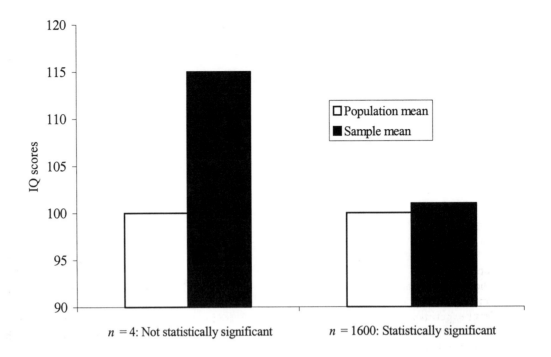

FIGURE 6.1 The influence of sample size on statistical significance.

We can examine the simplest form of effect size by returning to our examples using IQ scores. Remember that we have a population with an average IQ score of 100 and a standard deviation of 15. We also had two samples, each with average IQ scores of 105, and one with a sample size of 25 and the other with a sample size of 1,600. Also recall that to find the standard error for calculating our z scores, we simply divided the standard deviation by the square root of the sample size. So for the sample with 25 members, our standard error was

$$\sigma_{\bar{x}} = 15 \div \sqrt{25} \Rightarrow 15 \div 5 \Rightarrow 3$$

where $\sigma_{\bar{x}}$ is the standard error of the mean

To calculate an effect size, what we need to do is convert this standard error back into a standard deviation. If we *divide* the standard deviation by the square root of the sample size to find the standard error, we can *multiply* the standard error by the square root of the sample size to find the standard deviation. When we do this, we find that

$$\sigma = 3 * \sqrt{25} \Rightarrow 3 * 5 \Rightarrow 15$$

where σ is the population standard deviation

Notice that the standard deviation would be exactly the same if we calculated it for the larger sample size of 1,600, even though the standard error was much smaller for that sample.

$$\sigma = .375 * \sqrt{1600} \Rightarrow .375 * 40 \Rightarrow 15$$

Once we have our standard deviation, it is easy to calculate an effect size, which has they symbol *d*. In the IQ example, we could determine the effect size as follows:

$$d = \frac{105 - 100}{15}$$

$$d = \frac{5}{15}$$

$$d = .33$$

where *d* is the effect size

As you can see, the formula for the effect size translates the numerator into standard deviation units. When the numerator represents some sort of difference score (e.g., the difference between two or more group means, the difference between a sample statistic and a population parameter), the effect size will represent that difference in standard deviation units. This is similar to representing the difference in standard error units, as most inferential statistics do (e.g., *t* values, *F* values, correlation coefficients), except that sample size is *eliminated* from the process.

There are no hard and fast rules regarding the interpretation of effect sizes. Some textbook authors suggest that effect sizes smaller than .20 are small, those between .25 and .75 are moderate, and those over .80 are large. When determining whether an effect size is meaningful, it is important to consider what you are testing. If I am comparing the mortality rates of two samples trying two different experimental drugs, even small effect sizes are important, because we are talking about life and death. But if I'm comparing two different samples' preference for ice cream flavors, even fairly large effect sizes may have little more than trivial importance in the real world.[4]

The recent push by some researchers to focus more heavily on effect sizes than on statistical significance reminds me that I should conclude this section of the chapter by urging you to take both effect size *and* statistical significance into consideration as you read and conduct research. Notice that in the previous examples, the exact same effect size was produced with the 25-person sample as with the 1,600-person sample. These results suggest that sample size does not matter. In fact, sample size is very important. Stated simply, it is easier to come up with fluke, or *chance* results with smaller sample sizes than with larger sample sizes. Our tests of statistical significance, which are sensitive to sample size, tell us the probability that our results are just due to random sampling error, or chance. Because larger sample sizes have a better likelihood of representing the populations from which they were selected, the results of studies that use larger sample sizes are more *reliable* than those using smaller sample sizes, if all else is equal (e.g., how the samples were selected, the methods used in the study, etc.). When used together, tests of statistical significance and measures of effect size can provide important information regarding the reliability and importance of statistical results. Of course, our own judgments about the meaning, causes, and consequences of our results are also important factors.

EXAMPLE: STATISTICAL SIGNIFICANCE AND EFFECT SIZE OF GENDER DIFFERENCES IN MOTIVATION

To illustrate the concepts of statistical significance and effect size, I present the results from an independent samples *t* test that I conducted using data from research I conducted with high school students. In this study, 483 students (240 girls, 243 boys) were given surveys in their social studies classrooms to measure their motivation, beliefs, and attitudes about school and schoolwork. One of the constructs that my colleague and I measured was a motivational orientation called *performance-approach goals*. Performance-approach goals refer to students' perceptions that one purpose of trying to achieve academically is to demonstrate to others how smart they are, sometimes by outperforming other students. We used a measure of performance-approach goals that was

---

[4] There are other types of effect sizes frequently reported in research. One of the most common of these is the percentage of variance explained by the independent variable. I mention this in later chapters as I discuss the concept of explained variance.

developed by Carol Midgley and her colleagues at the University of Michigan (Midgley et al., 1998). This measure includes five items: (1) "I'd like to show my teacher that I'm smarter than the other students in this class"; (2) "I would feel successful in this class if I did better than most of the other students"; (3) "I want to do better than other students in this class"; (4) "Doing better than other students in this class is important to me"; and (5) "I would feel really good if I were the only one who could answer the teachers' questions in this class." Students responded to each of these questions using a 5-point Likert scale (with 1 = "not at all true" and 5 = "very true"). Students' responses to these five items were then averaged, creating a performance-approach goals scale with a range from 1 to 5.

I wanted to see whether boys and girls in my sample of high school students had different means on this performance-approach goals scale. Some scholars have argued that boys tend to have a more individualistic, competitive view of the world than do girls, so we might expect boys to have higher average scores on the performance-approach goals scale than will girls. Others think this is hogwash. So I used the SPSS statistics program to conduct an independent samples t test. In Table 6.1, the actual SPSS output from the t test is presented. In this output, we can see that the sample size, mean, standard deviation, and standard error of the mean are presented separately for the sample of boys and the sample of girls. Below this, the difference between the boys' and girls' means on the performance-approach goals scale is presented ($\overline{X}_{girls} - \overline{X}_{boys} = -.2175$). Below this mean difference, a t value is presented ($t = -2.46$), followed by the degrees of freedom ($N_1 + N_2 - 2 = 481$). Next, under the heading "2-Tail Sig" the p value is presented. This p value is the probability of obtaining a t value of this size just due to random sampling error, or chance. As we can see, this p value is quite small ($p = .014$). If we are using the conventional alpha level of .05 (i.e., $\alpha = .05$), then we can see that our p value is smaller than our alpha level, and we would reject the null hypothesis of no differences between boys' and girls' average scores on the performance-approach goals scale. Therefore, we would conclude that our results are statistically significant. The last piece of information provided in Table 6.1 is the standard error of the difference between the means, which equals .089 ($s_{\overline{x}1 - \overline{x}2} = .089$).

Using the information in Table 6.1, the information about conducting t tests presented in Chapter 8, and the table of t values presented in Appendix B, we can recreate this t test, see why it is statistically significant, and calculate the effect size. First, using the means and the standard error of the difference between the means, we can reproduce the equation used to generate the t value:

$$t = \frac{2.9377 - 3.1552}{.089}$$

$$t = \frac{-.2175}{.089}$$

$$t = -2.46$$

Next, using the degrees of freedom ($df = 481$), we can look in Appendix B to find the approximate probability of finding a t value of this size by chance. Because our degrees of freedom are larger than 120, we must look in the row labeled with the infinity symbol (∞). The absolute value of our observed t value is 2.46, which falls between the t values of 2.326 and 2.576 in the infinity row of the table. Looking at the values under the heading "α level for a two-tailed test" we can see these two t values (2.326 and 2.576) have probabilities of .02 and .01, respectively. Therefore, we know that our t value of -2.46 has a probability of occurring by chance between 1% and 2% of the time. Our SPSS output confirms this, placing the probability at a more precise number, $p = .014$. This would be considered statistically significant if we are using the conventional alpha level of .05.

TABLE 6.1   *t* test comparing high school boys' and girls' average scores on the performance-approach goals scale.

Variable	Number of Cases	Mean	SD	SE of Mean
Performance-approach goals				
girl	240	2.9377	.922	.057
boy	243	3.1552	1.023	.069

Mean Difference = -.2175

	t value	df	2-Tail Sig	SE of Diff
Equal	-2.46	481	.014	.089

Now that we know girls in our sample scored significantly lower than boys on the performance-approach goals scale, we need to calculate an effect size. Notice that our sample size of 483 is fairly large, and our difference between the means (.2175 on a 5-point scale) is fairly small. So even though our result is statistically significant, is it *practically* significant? To calculate an effect size for an independent samples *t test*, we use the following formula:

$$d = \frac{\overline{X}_1 - \overline{X}_2}{s}$$

where *d* is the effect size
X_1 and X_2 are the two sample means
s is the standard deviation

Although we have a standard deviation for each of the two samples in the study, we do not yet have a single standard deviation to use in our effect size formula. To find it, multiply the standard error by the square root of the sample size, as we did in our earlier example. It is slightly more tricky here, however, because we've got two sample sizes. In this example, the two sample sizes are roughly equal, so we just multiply the standard error of the difference between the means by the square root of the average of these two samples (241.5). If our samples were of quite different size, we would need to use a more complicated formula for determining the standard deviation.

$$S = \sqrt{241.5} \; * .089$$

$$S = 15.54 * .089$$

$$S = 1.383$$

where *S* is the standard deviation of the difference between two independent sample means

Now that we have our standard deviation, we can easily calculate the effect size:

$$d = \frac{2.9377 - 3.1552}{1.383}$$

$$d = \frac{-.2175}{1.383}$$

$$d = -.157 \approx |.16|$$

Our effect size of .16 (note that we always take the absolute value when reporting effect sizes) is quite small. When we combine the results of our analysis of statistical significance with our effect size results, we have a somewhat mixed picture. First, we discovered that boys scored statistically significantly higher on the performance-approach goals scale than did girls. However, our small effect size ($d = .16$) suggests that this difference is due primarily to the large sample in our study, and that the differences between boys and girls on this scale may not be particularly important or meaningful.

Wrapping Up and Looking Forward

For several decades, statistical significance has been the measuring stick used by social scientists to determine whether the results of their analyses were meaningful. But as we have seen in this chapter and in our discussion of standard errors in Chapter 5, tests of statistical significance are quite dependent on sample size. With large samples, even trivial effects are often statistically significant, whereas with small sample sizes, quite large effects may not reach statistical significance. Because of this, there has recently been an increasing appreciation of, and demand for, measures of practical significance as well. When determining the practical significance of your own results, or of those you encounter in published articles or books, you are well advised to consider all of the measures at your disposal. Is the result statistically significant? How large is the effect size? And, as you look at the effect in your data and place your data in the context of real-world relevance, use your judgment to decide whether you are talking about a meaningful or a trivial result. In the chapters to come, we will encounter several examples of inferential statistics. Use what you have learned in this chapter to determine whether the results presented should be considered practically significant.

Glossary of Terms and Symbols for Chapter 6

Alpha level: The *a priori* probability of falsely rejecting the null hypothesis that the researcher is willing to accept. It is used, in conjunction with the *p* value, to determine whether a sample statistic is statistically significant.

Alternative hypothesis: The alternative to the null hypothesis. Usually, it is the hypothesis that there is some effect present in the population (e.g., two population means are not equal, two variables are correlated, a sample mean is different from a population mean, etc.).

Descriptive statistics: Statistics that describe the characteristics of a given sample or population. These statistics are only meant to describe the characteristics of those from whom data were collected.

Effect size: A measure of the size of the effect observed in some statistic. It is a way of determining the practical significance of a statistic by reducing the impact of sample size.

Inferential statistics: Statistics generated from sample data that are used to make inferences about the characteristics of the population the sample is alleged to represent.

Null hypothesis: The hypothesis that there is no effect in the population (e.g., that two population means are not different from each other, that two variables are not correlated in the population).

Population: The group from which data are collected or a sample is selected. The population encompasses the entire group for which the data are alleged to apply.

Practical significance: A judgment about whether a statistic is relevant, or of any importance, in the "real world."

p value: The probability of obtaining a statistic of a given size from a sample of a given size by chance, or due to random error.

Random chance: The probability of a statistical event occurring due simply to random variations in the characteristics of samples of a given sizes selected randomly from a population.

Random sampling error: The error, or variation, associated with randomly selecting samples of a given size from the population.

Sample: An individual or group, selected from a population, from whom or which data are collected.

Statistical significance: When the probability of obtaining a statistic of a given size due strictly to random sampling error, or chance, is less than the selected alpha level, the result is said to be statistically significant. It also represents a rejection of the null hypothesis.

Type I error: Rejecting the null hypothesis when in fact the null hypothesis is true.

p	Symbol for p value, or probability.
α	Symbol for alpha level.
d	Symbol for effect size.
S	Symbol for the standard deviation used in the effect size formula.
∞	Symbol for infinity.
s_x	Symbol for the standard error calculated with the sample standard deviation.
σ_x	Symbol for the standard error calculated with the population standard deviation.
H_0	Symbol for the null hypothesis.
H_A or H_1	Symbols for the alternative hypothesis

REFERENCES AND RECOMMENDED READING

Midgley, C., Kaplan, A., Middleton, M., Maehr, M. L., Urdan, T., Anderman, L. H., Anderman, E., & Roeser, R. (1998). The development and validation of scales assessing students' achievement goal orientations. *Contemporary Educational Psychology, 23*, 113-131.

CHAPTER 7

CORRELATION

In several of the previous chapters we have examined statistics and parameters that describe a single variable at a time, such as the mean, standard deviation, z scores, and standard errors. Although such single-variable statistics are important, researchers are often interested in examining the relations among two or more variables. One of the most basic measures of the association among variables, and a foundational statistic for several more complex statistics, is the **correlation coefficient**. Although there are a number of different types of correlation coefficients, the most commonly used in social science research is the **Pearson product-moment correlation coefficient**. Most of this chapter is devoted to understanding this statistic, with a brief description of three other types of correlations: the **point-biserial coefficient,** the **Spearman rho coefficient,** and the **phi coefficient.**

When to Use Correlation and What It Tells Us

Researchers compute correlation coefficients when they want to know how two variables are related to each other. For a Pearson product-moment correlation, both of the variables must be measured on an interval or ratio scale, otherwise known as **continuous variables**. For example, suppose I wanted to know whether there was a relationship between the amount of time students spend studying for an exam and their scores on the exam. I suspect that the more hours students spend studying, the higher their scores will be on the exam. But I also suspect that there is not a perfect correspondence between time spent studying and test scores. Some students will probably get low scores on the exam even if they study for a long time, simply because they may have a hard time understanding the material. Indeed, there will probably be a number of students who spend an inordinately long period of time studying for the test precisely *because* they are having trouble understanding the material. On the other hand, there will probably be some students who do very well on the test without spending very much time studying. Despite these "exceptions" to my rule, I will still hypothesize that, *on average*, as the amount of time spent studying increases, so do students' scores on the exam.

There are two fundamental characteristics of correlation coefficients researchers care about. The first of these is the **direction** of the correlation coefficient. Correlation coefficients can be either positive or negative. A **positive correlation** indicates that the values on the two variables being analyzed move in the same direction. That is, as scores on one variable go up, scores on the other variable go up as well (on average). Similarly, on average, as scores on one variable go down, scores on the other variable go down. Returning to my earlier example, if there is a positive correlation between the amount of time students spend studying and their test scores, I can tell that, on average, the more time students spent studying, the higher their scores on the test. This is *equivalent* to saying that, on average, the *less* time they spent studying, the *lower* their scores on the test. Both of these represent a *positive* correlation between time spent studying and test scores. (**Note:** I keep saying "on average" because it is important to note that the presence of a correlation between two variables does <u>not</u> mean that this relationship holds true for each member of the sample or population. Rather, it means that, in general, there is a relationship of a given direction and strength between two variables in the sample or population.)

A **negative correlation** indicates that the values on the two variables being analyzed move in *opposite* directions. That is, as scores on one variable go up, scores on the other variable go down, and vice-versa (on average). If there were a negative correlation between the amount of time spent studying and test scores, I would know that, on average, the more time students spend studying for the exam, the *lower* they actually score on the exam. Similarly, with a negative correlation I would also conclude that, on average, the *less* time students spent studying, the *higher* their scores on the exam. These positive and negative correlations are represented by **scattergrams** in Figure 7.1. Scattergrams are simply graphs that indicate the scores of each case in a sample simultaneously on two variables. For example, in the "Positive Correlation" scattergram in Figure 7.1, the first case in the sample studied for 1 hour and got a score of 30 on the exam. The second case studied for 2 hours and scored a 40 on the exam.

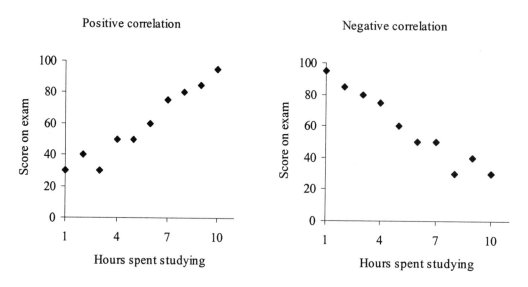

FIGURE 7.1 Examples of positive and negative correlations.

The second fundamental characteristic of correlation coefficients is the **strength** or **magnitude** of the relationship. Correlation coefficients range in strength from –1.00 to +1.00. A correlation coefficient of .00 indicates that there is no relationship between the two variables being examined. That is, scores on one of the variables are not related in any meaningful way to scores on the second variable. The closer the correlation coefficient is to either –1.00 or +1.00, the stronger the relationship is between the two variables. A **perfect negative correlation** of –1.00 indicates that for every member of the sample or population, a higher score on one variable is related to a lower score on the other variable. A **perfect positive correlation** of +1.00 reveals that for every member of the sample or population, a higher score on one variable is related to higher score on the other variable.

Perfect correlations are never found in actual social science research. Generally, correlation coefficients stay between -.70 and +.70. Some textbook authors suggest that correlation coefficients between -.20 and +.20 indicate a weak relation between two variables, those between .20 and .50 (either positive or negative) represent a moderate relationship, and those larger than .50 (either positive or negative) represent a strong relationship. These general rules of thumb for judging the relevance of correlation coefficients must be taken with a grain of salt. For example, even a "small" correlation between alcohol consumption and liver disease (e.g., +.15) is important, whereas a strong correlation between how much children like vanilla and chocolate ice cream (e.g., +.70) may not be so important.

The scattergrams presented in Figure 1 represent very strong positive and negative correlations (*r* = **.97** and *r* = **-.97** for the positive and negative correlations, respectively; *r* is the symbol for the sample Pearson correlation coefficient). In Figure 7.2, a scattergram representing virtually no correlation between the number of hours spent studying and the scores on the exam is presented. Notice that there is no discernible pattern between the scores on the two variables. In other words, the data presented in Figure 7.2 reveal that it would be virtually impossible to predict an individual's test score simply by knowing how many hours the person studied for the exam.

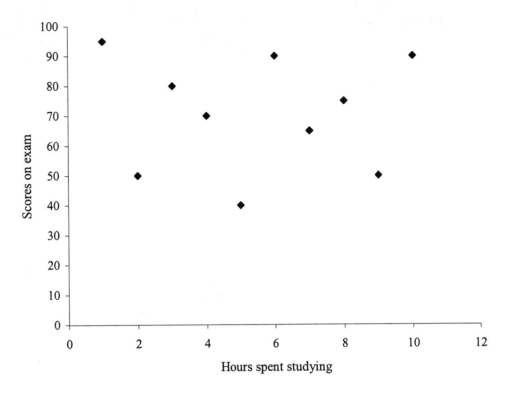

FIGURE 7.2   No correlation between hours spent studying and exam scores

PEARSON CORRELATION COEFFICIENTS IN DEPTH

The first step in understanding how Pearson correlation coefficients are calculated is to notice that we are concerned with a sample's scores on *two* variables at the same time. Returning to our previous example of study time and test scores, suppose that we randomly select a sample of five students and measure the time they spent studying for the exam and their exam scores. The data are presented in Table 7.1.

TABLE 7.1   Data for correlation coefficient.

	Hours Spent Studying (X Variable)	Exam Score (Y Variable)
Student 1	5	80
Student 2	6	85
Student 3	7	70
Student 4	8	90
Student 5	9	85

For these data to be used in a correlation analysis, it is critical that the scores on the two variables are *paired*. That is, for each student in my sample, the score on the X variable (hours spent studying) is paired with his or her own score on the Y variable (exam score). If I want to determine the relation between hours spent studying and exam scores, I cannot pair Student 1's hours spent studying with Student 4's test score. I must match each student's score on the X variable with his or her own score on the Y variable. Once I have done this, I can determine whether, on average, hours spent studying is related to exam scores.

Calculating the Correlation Coefficient

There are several different formulas that can be used to calculate Pearson product-moment correlation coefficients. Each of these formulas produces the same result and differ only in their ease of use. In fact, none of them are particularly easy to use. I present one of them here to illustrate the principles of correlation coefficients. Should you find yourself in need of knowing the relation between two variables, I recommend that you use a calculator or statistics program that generates correlation coefficients for you.

The formula presented in Table 7.2 requires that you *standardize* your variables (see Chapter 4). Remember that when you standardize a variable, you are simply subtracting the mean from each score in your sample and dividing by the standard deviation. What this does is provide a **z score** for each case in the sample. Those members of the sample with scores below the mean will have negative z scores, whereas those members of the sample with scores above the mean will have positive z scores.

TABLE 7.2   Definitional formula for Pearson correlation.

$$r = \frac{\Sigma(z_x z_y)}{N}$$

where r = Pearson product-moment correlation coefficient
z_x = a z score for variable X
z_y = a paired z score for variable Y
N = the number of pairs of X and Y scores

Notice that this formula looks similar to some of the other formulas that we have already encountered. For example, the denominator is N, which is the number of pairs of scores (i.e., the number of cases in the sample). Whenever we divide by N, we are finding an *average*. This was true when we examined the formula for the mean in Chapter 1 and the formulas for variance and standard deviation in Chapter 2. So we know that the correlation coefficient will be an *average* of some kind. But what is it an average of? Now take a look at the numerator. Here, we see that we must find the sum (Σ) of something. Recall that when discussing the formulas for the variance and standard deviation in Chapter 2, we also encountered this summation sign in the numerator. There, we had to find the sum of the squared deviations between each individual score and the mean. But in the formula for computing the correlation coefficient, we have to find the sum of the **cross products** between the z scores on each of the two variables being examined for each case in the sample. When we multiply each individual's score on one variable with that individual's score on the second variable (i.e., find a cross product), sum those across all of the individuals in the sample, and then divide by N, we have an average cross product, and this is known as **covariance**. If we standardize this covariance, we end up with a correlation coefficient. In the formula provided in Table 7.2, we simply standardized the variables before we computed the cross products, thereby producing a standardized covariance statistic, which is a correlation coefficient.

In this formula, notice what is happening. First, we are multiplying the paired z scores together. When we do this, notice that if an individual case in the sample has scores above the mean on each of the two variables being examined, the two z scores being multiplied will both be positive, and the resulting cross product will also be positive. Similarly, if an individual case has scores below the mean on each of the two variables, the two z scores being multiplied will both be negative, and the cross product will again be positive. Therefore, if we have a sample where low scores on one variable tend to be associated with low scores on the other variable, and high scores on one variable tend to be associated with high scores on the second variable, then when we add up to products from our multiplications we will end up with a *positive* number. This is how we get a positive correlation coefficient.

Now consider what happens when high scores on one variable are associated with low scores on the second variable. If an individual case in a sample has a score that is higher than the average on the first variable (i.e., a positive z score) and a score that is below the mean on the second variable (i.e., a negative z score), when these two z scores are multiplied together, they will produce a *negative* product. If, for most of the cases in the sample, high scores on one variable are associated with low scores on the second variable, the sum of the products of the z scores $[\Sigma(z_x z_y)]$ will be *negative*. This is how we get a negative correlation coefficient.

What the Correlation Coefficient Does, and Does Not, Tell Us

Correlation coefficients such as the Pearson are very powerful statistics. They allow us to determine whether, on average, the values on one variable are *associated with* the values on a second variable. This can be very useful information, but people, including social scientists, are often tempted to ascribe more meaning to correlation coefficients than they deserve. Namely, people often confuse the concepts of *correlation* and **causation**. Correlation (co-relation) simply means that variation in the scores on one variable *correspond* with variation in the scores on a second variable. Causation means that variation in the scores on one variable *cause* or *create* variation in the scores on a second variable.

When we make the leap from correlation to causation, we may be wrong. As an example, I offer this story, which I heard in my introductory psychology class. As the story goes, one winter shortly after World War II, there was an explosion in the number of storks nesting in some northern European country (I cannot remember which). Approximately 9 months later, there was a large jump in the number of babies that were born. Now, the link between storks and babies being what it is, many concluded that this correlation between the number of storks and the number of babies represented a causal relationship. Fortunately, science tells us that babies do not come from storks after all, at least not human babies. However, there is something that storks and babies have in common: Both can be "summoned" by cold temperatures and warm fireplaces. It seems that storks like to nest in warm chimneys during cold winters. As it happens, cold winter nights also foster baby-making behavior. The apparent cause-and-effect relationship between storks and babies was in fact caused by a third variable: a cold winter.

For a more serious example, we can look at the relationship between SAT scores and first-year college grade point average. The correlation between these two variables is about .40. Although these two variables are moderately correlated, it would be difficult to argue that higher SAT scores *cause* higher achievement in the first year of college. Rather, there is probably some other variable, or set of variables, responsible for this relationship. For example, we know that taking a greater number of advanced math courses in high school is associated with higher SAT scores *and* with higher grades in first-year math courses in college.

The point of these examples is simple: Evidence of a relationship between two variables (i.e., a correlation) does not necessarily mean that there is a causal relationship between the two variables. However, it should also be noted that a correlation between two variables is a *necessary ingredient* of any argument that the two variables are causally related. In other words, I cannot claim that one variable causes another (e.g., that smoking causes cancer) if there is no correlation between smoking and cancer. If I do find a correlation between smoking and cancer, I must rule out other factors before I can conclude that it is smoking that causes cancer.

In addition to the correlation-causation issue, there are a few other important features of correlations worth noting. First, simple Pearson correlations are designed to examine *linear* relations among variables. In other words, they describe average *straight* relations among variables. For example, if you find a positive correlation between two variables, you can predict how much the scores in one variable will increase with each corresponding increase in the second variable. But not all relations between variables are linear. For example, there is a **curvilinear** relationship between anxiety and performance on a number of academic and nonacademic behaviors. When taking a math test, for example, a little bit of anxiety actually may help performance. However, once a student becomes too nervous, this anxiety can interfere with performance. We call this a curvilinear relationship because what began as a positive relationship between performance and anxiety at lower levels of anxiety becomes a negative relationship at higher levels of anxiety. This curvilinear relationship is presented graphically in Figure 7.3. Because correlation coefficients show the *average* relation between two variables, when the relationship between two variables is curvilinear, the correlation coefficient can be quite small, suggesting a weaker relationship than may actually exist.

TABLE 7.3 Data for studying-exam score correlation.

	Hours Spent Studying (X Variable)	Exam Score (Y Variable)
Student 1	0	95
Student 2	2	95
Student 3	4	100
Student 4	7	95
Student 5	10	100

Another common problem that arises when examining correlation coefficients is the problem of **truncated range**. This problem is encountered when the scores on one or both of the variables in the analysis do not have much variety in the distribution of scores, possibly due to a ceiling or floor effect. For example, suppose that I gave a sample of students a very easy test with a possible high score of 100. Then suppose I wanted to see if there was a correlation between scores on my test and how much time students spent studying for the test. Suppose I got the following data, presented in Table 7.3.

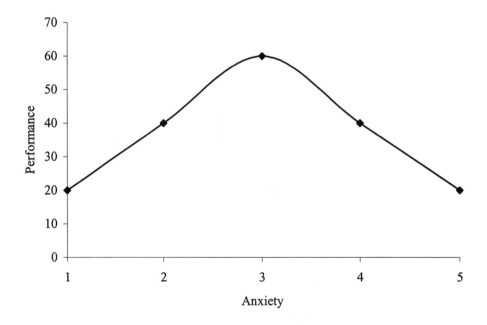

FIGURE 7.3 A curvilinear relationship.

In this example, all of my students did well on the test, whether they spent many hours studying for it or not. Because the test was too easy, a ceiling effect may have occurred, thereby truncating the range of scores on the exam. Although there may be a relationship between how much time students spend studying and their knowledge of the material, my test was not sensitive enough to reveal this relationship. The weak correlation that will be produced by the data in Table 7.3 may not reflect the true relationship between how much students study and how much they learn.

Statistically Significant Correlations

When researchers calculate correlation coefficients, they often want to know whether a correlation found in sample data represents the existence of a relationship between two variables in the population from which the sample was selected. In other words, they want to test whether the correlation coefficient is statistically significant (see Chapter 6 for a discussion of statistical significance). To test whether a correlation coefficient is statistically significant, the researcher begins with the null hypothesis that there is absolutely no relationship between the two variables in the population, or that the correlation coefficient in the population equals zero. The alternative hypothesis is that there, in fact, is a statistical relationship between the two variables in the population, and that the population correlation coefficient is *not* equal to zero. So what we are testing here is whether our correlation coefficient is statistically significantly different from 0. These two competing hypothesis can be expressed with symbols:

$$H_o: \rho = 0$$

$$H_1: \rho \neq 0$$

where ρ is rho, the population correlation coefficient.

The t distribution is used to test whether a correlation coefficient is statistically significant. Therefore, we must conduct a t test.   As with all t tests, the t test that we use for correlation coefficients is involves a ratio, or fraction. The numerator of the fraction is the difference between two values. The denominator is the standard error. When we want to see whether a sample correlation coefficient is significantly different from zero, the numerator of the t test formula will be the sample correlation coefficient, r, minus the hypothesized value of the population correlation coefficient (ρ), which in our null hypothesis is zero. The denominator will be the standard error of the sample correlation coefficient:

$$t = \frac{r - \rho}{s_r}$$

where r is the sample correlation coefficient,
ρ is the population correlation coefficient,
s_r is the standard error of the standard error of the sample correlation coefficient.

Fortunately, with the help of a little algebra, we do not need to calculate s_r to calculate the t value for correlation coefficients. First, for the sake of knowledge, let me present the formula for s_r :

$$s_r = \sqrt{(1 - r^2)(N - 2)}$$

where r^2 is the correlation coefficient squared and
N is the number of cases in the sample.

The formula for calculating the t value is

$$t = (r)\sqrt{\frac{N - 2}{1 - r^2}}$$

where **degrees of freedom** is the number of cases in the sample minus two ($df = N - 2$)

To illustrate this formula in action, let's consider an example. Some research suggests that there is a relationship between the number of hours of sunlight people are exposed to during the day and their mood. People living at extreme northern latitudes, for example, are exposed to very little sunlight in the depths of winter days and weeks without more than a few hours of sunlight per day. There is some evidence that such sunlight deprivation is related to feelings of depression and sadness. In fact, there is even a name for the condition: seasonal affective disorder, or SAD. To examine this relationship for myself, I randomly select 100 people from various regions of the world, measure the time from sunrise to sunset on a given day where each person lives, and get a measure of each person's mood on a scale from 1 to 10 (1 = "very sad," 10 = "very happy"). Because the members of my sample live at various latitudes, the number of daylight hours will vary. If I conduct my study in January, those participants living in the north will have relatively short days whereas those living in the south will have long days.

Suppose that I were to compute a Pearson correlation coefficient with these data and found that the correlation between number of sunlight hours in the day and scores on the mood scale was r = .25. Is this a statistically significant correlation? To answer that question, we must find a t value associated with this correlation coefficient and determine that probability of obtaining a t value of this size by chance (see Chapter 8). In this example,

$$t = (.25)\sqrt{\frac{100-2}{1-.25^2}}$$

$$t = (.25)\sqrt{\frac{98}{1-.25^2}}$$

$$t = (.25)\sqrt{\frac{98}{1-.0625}}$$

$$t = (.25)\sqrt{\frac{98}{1-.9375}}$$

$$t = (.25)\sqrt{104.53}$$

$$t = (.25)10.22$$

$$t = 2.56, df = 98$$

To see whether this t value is statistically significant, we must look at the table of t values in Appendix B. There we can see that, because our degrees of freedom equals 98, we must look at t values in both the $df = 60$ row and the $df = 120$ row. Looking at the $df = 60$ row, we can see that a t value of 2.56 has a probability of between .01 and .02 (for a two-tailed test). We get the same results when looking in the $df = 120$ row. Therefore, we conclude that our p value is between .01 and .02. If our alpha level is the traditional .05, we would conclude that our correlation coefficient is statistically significant. In other words, we would conclude that, on the basis of our sample statistic, in the larger population of adults the longer the daylight hours, the better their mood, on average. We could convey all of that information to the informed reader of statistics by writing "We found a significant relationship between number of daylight hours and mood ($r = .25$, $t_{(98)} = 2.56$, $p < .05$)."

This example also provides a good opportunity to once again remind you of the dangers of assuming that a correlation represents a causal relationship between two variables. Although it may well be that longer days cause the average adult to feel better, these data do not prove it. An alternative causal explanation for our results is that shorter days are also associated with *colder* days whereas longer days are generally associated with *warmer* days. It may be the case that *warmth* causes better moods whereas the lack of warmth causes depression and sadness. If people had warm, short days they might be just as happy as if they had warm, long days. So remember: Just because two variables are correlated, it does not mean that one causes the other.

The Coefficient of Determination

Although correlation coefficients give an idea of the strength of the relationship between two variables, they often feel a bit nebulous. If you get a correlation coefficient of .40, is that a strong relationship? Fortunately, correlation coefficients can be used to obtain a more concrete-feeling statistic: the **coefficient of determination**. Even better, it is easy to calculate.

When we want to know if two variables are related to each other, we are really asking a somewhat more complex question: Are the variations in the scores on one variable somehow associated with the variations in the scores on a second variable? Put another way, a correlation coefficient tells us whether we can know anything about the scores on one variable if we already know the scores on a second variable. In common statistical vernacular, what we want to be able to do with a measure of association, like a correlation coefficient, is be able to *explain* some of the variance in the scores on one variable with the scores on a second variable. The coefficient of determination tells us how much of the variance in the scores of one variable can be understood, or explained, by the scores on a second variable.

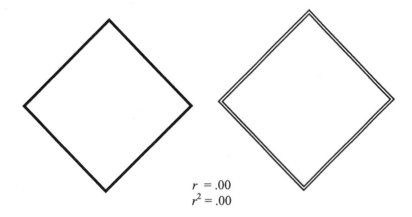

$r = .00$
$r^2 = .00$

FIGURE 7.4 Uncorrelated variables.

One way to conceptualize **explained variance** is to understand that when two variables are correlated with each other, they *share* a certain percentage of their variance. Consider an example. If we have a sample of 10 people, and we measure their height and their weight, we've got 10 scores on each of two variables. Assuming that my 10 people differ in how tall they are, there will be some total amount of variance in their scores on the height variable. There will also be some total amount of variance in their scores on the weight variable, assuming that they do not all weigh the same amount. These total variances are depicted in Figure 7.4 as two full squares, each representing 100% of the variance in their respective variables. Notice how they do not overlap.

When two variables are related, or correlated, with each other, there is a certain amount of **shared variance** between them. In Figure 7.4, the two squares are not touching each other, suggesting that all of the variance in each variable is independent of the other variable. There is no overlap. But when two variables are correlated, there is some *shared* variance. The stronger the correlation, the greater the amount of shared variance, and the more variance you can explain in one variable by knowing the scores on the second variable. The precise percentage of shared, or explained, variance can be determined by squaring the correlation coefficient. This squared correlation coefficient is known as the coefficient of determination. Some examples of different coefficients of determination are presented in Figure 7.5. The stronger the correlation, the greater the amount of shared variance, and the higher the coefficient of determination. It is still important to remember that even though the coefficient of determination is used to tell us how much of the variance in one variable can be explained by the variance in a second variable, coefficients of determination do not necessarily indicate a *causal* relationship between the two variables.

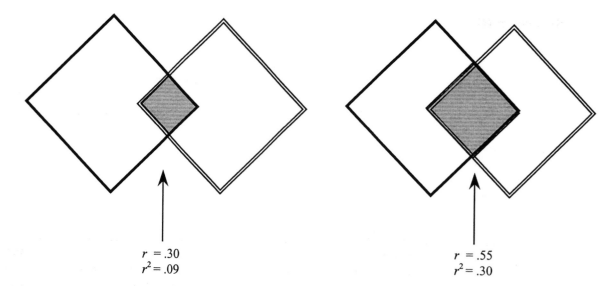

FIGURE 7.5 Examples of different coefficients of determination.

A BRIEF WORD ON OTHER TYPES OF CORRELATION COEFFICIENTS

Although Pearson correlation coefficients are probably the most commonly used and reported in the social sciences, they are limited by the requirement that both variables are measured on interval or ratio scales. Fortunately, there are methods available for calculating the strength of the relationship between two variables even if one or both variables are not measured using interval or ratio scales. In this section I briefly describe three of these "other" correlation coefficients. It is important to note that all of these statistics are very similar to the Pearson correlation coefficient and each produces a correlation coefficient that is similar to the Pearson *r*. They are simply specialized versions of the Pearson correlation coefficient that can be used when one or both of the variables are not measured using interval or ratio scales.

Point Biserial

When one of our variables is a continuous variable (i.e., measured on an interval or ratio scale) and the other is a two-level categorical (a.k.a. nominal) variable (also known as a **dichotomous variable**), we need to calculate a point-biserial correlation coefficient. This coefficient is a specialized version of the Pearson correlation coefficient discussed earlier in this chapter. For example, supposed I wanted to know whether there was a relationship between whether a person owns a car (yes or no) and their score on a written test of traffic rule knowledge, such as the tests one must pass to get a driver's license. In this example, we are examining the relation between one categorical variable with two categories (whether one owns a car) and one continuous variable (one's score on the driver's test). Therefore, the point-biserial correlation is the appropriate statistic in this instance.

Phi

Sometimes researchers want to know whether two dichotomous variables are correlated. In this case, we would calculate a phi coefficient (Φ), which is another specialized version of the Pearson *r*. For example, suppose I wanted to know whether gender (male, female) was associated with whether one smokes cigarettes or not (smoker, non smoker). In this case, with two dichotomous variables, I would calculate a phi coefficient. (**Note:** Those readers familiar with chi-square analysis will notice that two dichotomous variables can also be analyzed using chi-square.)

Spearman Rho

Sometimes data are recorded as ranks. Because ranks are a form of ordinal data, and the other correlation coefficients discussed so far involve either continuous (interval, ratio) or dichotomous variables, we need a different type of statistic to calculate the correlation between two variables that use ranked data. The **Spearman rho**, a specialized form of the Pearson *r*, is appropriate. For example, many schools use students' grade point averages (a continuous scale) to rank students (an ordinal scale). In addition, students' scores on standardized achievement tests can be ranked. To see whether a students' rank in their school is related to their rank on the standardized test, a Spearman rho coefficient can be calculated.

EXAMPLE: THE CORRELATION BETWEEN GRADES AND TEST SCORES

Student achievement can be measured in a variety of ways. One common method of evaluating students is to assign them letter grades. These grades can be converted into numbers (e.g., A = 4, B = 3, etc.). In high school, students grades across all of their classes (e.g., mathematics, science, social studies, etc.) can be combined into an average, thereby creating a grade point average (GPA), which is measured on a continuous, interval scale ranging from a possible low of 0 to a possible high of 4.33 (if the school gives grades of A+). Because grades are assigned by teachers, they are sometimes considered to be overly subjective. That is, different teachers may assign different grades to the same work. Similarly, there are some individual teachers that may give different grades to two students who produce the same quality of work. To ostensibly overcome such subjectivity, another form of assessment, the standardized test, was created. With this type of assessment, all students of a given grade level answer the same questions and their responses are scored by computer, thereby removing the human element and its subjectivity.

Some argue that standardized tests of ability and teacher-assigned grades generally measure the same thing. That is, for the most part, bright students will score well both on the test and in their grades. Others argue that standardized tests of ability and teacher-assigned grades really measure somewhat different things. Whereas standardized tests may measure how well students answer multiple-choice questions, teachers have the benefit of knowing students, and can take things like students' effort, creativity, and motivation into account when assigning grades. The first step in discovering which of these two viewpoints is more accurate is to see how strongly grades and test scores are related to each other. If there is a very strong correlation between the two, then both grades and test scores may in fact be measuring the same general trait. But if the two scores are only moderately correlated, perhaps they really do measure separate constructs. By constructs, I mean the actual "thing" that we are trying to measure. In the preceding example, if grades and test scores are strongly correlated, we could argue that both of these measures represent some underlying construct, such as "intelligence" or "academic ability." On the other hand, if these two variables are not strongly correlated, they each may represent different things, or constructs.

My colleague, Carol Giancarlo, and I recently collected data from a sample of 314 eleventh-grade students at a high school in California. Among the data we collected were their cumulative GPA (i.e., their GPA accumulated from the time they began high school until the time the data were collected). In addition, we gave students the Naglieri Nonverbal Test of Abilities (NNTA; Naglieri, 1996), a nonverbal test of general mental reasoning and critical thinking skills. To see if there was a statistically significant correlation between these two measures of ability, I used the SPSS statistical software program to calculate a correlation coefficient and a *p* value. The SPSS printout from this analysis is presented in Table 7.5.

The results presented in Table 7.5 provide several pieces of information. First, there are three correlation coefficients presented. The correlations on the diagonal show the correlation between a single variable and itself. Therefore, the first correlation coefficient presented reveals that GPA is correlated with itself perfectly ($r = 1.0000$). Because we always get a correlation of 1.00

when we correlate a variable with itself, these correlations presented on the diagonal are meaningless. That is why there is not a *p* value reported for them. The numbers in the parentheses, just below the correlation coefficients, report the sample size. There were 314 eleventh-grade students in this sample. The correlation coefficient that is *off* the diagonal is the one we're interested in. Here, we can see that students' GPA was moderately correlated with their scores on the Naglieri test ($r = .4291$). This correlation is statistically significant, with a *p* value of less than .0001 ($p < .0001$).

TABLE 7.5   SPSS printout of correlation analysis.

	GPA	Naglieri
GPA	1.0000 (314) $p = .$	
Naglieri	.4291 (314) $p = .000$	1.0000 (314) $p = .$

To gain a clearer understanding of the relationship between GPA and Naglieri test scores, we can calculate a coefficient of determination. We do this by squaring the correlation coefficient. When we square this correlation coefficient ($.4291 * .4291 = .1841$), we see that GPA explains a little bit more than 18% of the variance in the Naglieri test scores. Although this is a substantial percentage, it still leaves more than 80% of the ability test scores unexplained. Because of this large percentage of unexplained variance, we must conclude that teacher-assigned grades reflect something substantially different from general mental reasoning abilities and critical thinking skills, as measured by the Naglieri test.

WRAPPING UP AND LOOKING FORWARD

Correlation coefficients, in particular Pearson correlation coefficients, provide a way to determine both the direction and the strength of the relationship between two variables measured on a continuous scale. This index can provide evidence that two variables are related to each other, or that they are not, but does not, in and of itself, demonstrate a *causal* association between two variables. In this chapter, we also were introduced to the concepts of *explained* or *shared variance* and the coefficient of determination. Determining how much variance in one variable is shared with, or explained by, another variable is at the core of all of the statistics that are discussed in the remaining chapters of this book. In particular, correlation coefficients are the precursors to the more sophisticated statistics involved in multiple regression (Chapter 12). In the next chapter, we examine *t* tests, which allow us to look at the association between a two-category independent variable and a continuous dependent variable.

GLOSSARY OF TERMS AND SYMBOLS FOR CHAPTER 7

Causation: The concept that variation in one variable *causes* variation in another variable.
Coefficient of determination: A statistic found by squaring the Pearson correlation coefficient that reveals the percentage of variance explained in each of the two correlated variables by the other variable.
Continuous variables: Variables that are measured using an interval or ratio scale.
Correlation coefficient: A statistic that reveals the strength and direction of the relationship between two variables.

Curvilinear: A relationship between two variables that is positive at some values but negative at other values.

Degrees of freedom: Roughly, the minimum amount of data needed to calculate a statistic. More practically, it is a number, or numbers, used to approximate the number of observations in the data set for the purpose of determining statistical significance.

Dichotomous variable: A categorical, or nominal, variable with two categories.

Direction: A characteristic of a correlation that describes whether two variables are positively or negatively related to each other.

Explained variance: The percentage of variance in one variable that we can account for, or understand, by knowing the value of the second variable in the correlation.

Negative correlation: A descriptive feature of a correlation indicating that as scores on one of the correlated variables increase scores on the other variable decrease, and vice-versa.

Pearson product-moment correlation coefficient: A statistic indicating the strength and direction of the relation between two continuous variables.

Perfect negative correlation: A correlation coefficient of $r = -1.0$. When the increasing scores of a given size on one of the variables in a correlation are associated with decreasing scores of a related size on the second variable in the correlation (e.g., for each 1-unit increase in the score on variable X there is a corresponding 2-unit decrease in the scores on variable Y).

Perfect positive correlation: A correlation coefficient of $r = +1.0$. When the increasing scores of a given size on one of the variables in a correlation are associated with increasing scores of a related size on the second variable in the correlation (e.g., for each 1-unit increase in the score on variable X there is a corresponding 2-unit increase in the scores on variable Y).

Phi coefficient: The coefficient describing the correlation between two dichotomous variables.

Point-biserial coefficient: The coefficient describing the relationship between one interval or ratio scaled (i.e., continuous) variable and one dichotomous variable.

Positive correlation: A characteristic of a correlation; when the scores on the two correlated variables move in the same direction, on average. As the scores on one variable rise, scores on the other variable rise, and vice-versa.

Raw-score formula: A formula used to calculate a statistic using only raw scores.

Shared variance: The concept of two variables overlapping such that some of the variance in each variable is shared. The stronger the correlation between two variables, the greater the amount of shared variance between them.

Spearman rho coefficient: The correlation coefficient used to measure the association between two variables measured on an ordinal scale (e.g., ranked data).

Strength, magnitude: A characteristic of the correlation with a focus on how strongly two variables are related.

Truncated range: When the responses on a variable are clustered near the top of the bottom of the possible range of scores, thereby limiting the range of scores and possibly limiting the strength of the correlation.

z score: Standardized score.

r	Symbol for the sample Pearson correlation coefficient.
ρ	Symbol for rho, the population correlation coefficient.
s_r	Symbol for the standard error of the correlation coefficient.
r^2	Symbol for the coefficient of determination.
df	Symbol for degrees of freedom.
Φ	Symbol for the phi coefficient, which is the correlation between two dichotomous variables.

REFERENCES AND RECOMMENDED READING

Hinkle, D. E., Wiersma, W., & Jurs, S. G. (1998). *Applied statistics for the behavioral sciences* (4th Ed.). Boston: Houghton Mifflin.

Naglieri (1996). *The Naglieri nonverbal ability test*. San Antonio, TX: Harcourt Brace.

CHAPTER 8

t TESTS

What Is a *t* Test?

Because there is a distinction between the common statistical vernacular definition of *t* tests and the more technical definition, *t* tests can be a little confusing. The common-use definition or description of *t* tests is simply comparing two means to see if they are significantly different from each other. The more technical definition or description of a *t* test is any statistical test that uses the *t*, or Student's *t*, family of distributions. In this chapter, I will briefly describe the family of distributions known as the *t* distribution. Then I will discuss the two most commonly conducted *t* tests, the **independent samples *t* test** and the **paired or dependent samples *t* test**.

t Distributions

In Chapters 3 and 4, I discussed the normal distribution and how to use the normal distribution to find *z* scores. The probabilities that are based on the normal distribution are accurate when (a) the population standard deviation is known, and/or (b) we have a large sample (i.e., $n > 120$). If neither of these is true, then we cannot assume that we have a nicely shaped bell curve and we cannot use the probabilities that are based on this normal distribution. Instead, we have to adjust our probability estimates by taking our sample size into account. As I discussed in Chapter 5, we are fortunate to have a set of distributions that have already been created for us that do this, and this is known as the family of *t* distributions. Which specific *t* distribution you use for a given problem depends on the size of your sample. There is a table of probabilities based on the different *t* distributions in Appendix B.

The Independent Samples *t* Test

One of the most commonly used *t* tests is the independent samples *t* test. You use this test when you want to compare the means of two *independent* samples on a given variable. For example, if you wanted to compare the average height of 50 randomly selected men to that of 50 randomly selected women, you would conduct an independent samples *t* test. Note that the sample of men is not related to the sample of women, and there is no overlap between these two samples (i.e., one cannot be a member of both groups). Therefore, these groups are *independent*, and an independent samples *t* test is appropriate. To conduct an independent samples *t* test, you need one **categorical** or **nominal independent variable** and one **continuous** or **intervally scaled dependent variable**. An dependent variable is a variable on which the scores may differ, or *depend* on the value of the independent variable. An independent variable is the variable that may cause, or simply be used to predict, the value of the dependent variable. The independent variable in a *t* test is simply a variable with two categories (e.g., men and women, fifth graders and ninth graders, etc.). In this type of *t* test, we want to know whether the average scores on the dependent variable differ according to which group one belongs to (i.e., the level of the independent variable). For example, we may want to know if the average height of people (height is the dependent, continuous variable) *depends* on whether the person is a man or a woman (gender of the person is the independent, categorical variable).

Dependent Samples *t* Test

A dependent samples *t* test is also used to compare two means on a single dependent variable. Unlike the independent samples test, however, a dependent samples *t* test is used to compare the means of a single sample or of two **matched** or **paired samples**. For example, if a group of students took a math test in March and that same group of students took the same math test two months later in May, we could compare their average scores on the two test dates using a dependent samples *t* test. Or, suppose that we wanted to compare a sample of boys' Scholastic Aptitude Test (SAT) scores with their fathers' SAT scores. In this example, each boy in our study would be matched with his father. In both of these examples, each score is matched, or paired with, a second score. Because of this pairing, we say that the scores are *dependent* upon each other, and a dependent samples *t* test is warranted.

INDEPENDENT SAMPLES *t* TESTS IN DEPTH

To understand how *t* tests work, it may be most helpful to first try to understand the conceptual issues and then move to the more mechanical issues involved in the formulas. Because the independent and dependent forms of the *t* tests are quite different, I discuss each of them separately. Let's begin with the independent samples *t* test.

Conceptual Issues With the Independent Samples *t* Test

The most complicated conceptual issue in the independent samples *t* test involves the standard error for the test. If you think about what this *t* test does, you can see that it is designed to answer a fairly straightforward question: Do two independent samples differ from each other *significantly* in their average scores on some variable? Using an example to clarify this question, we might want to know whether a random sample of 50 men differs *significantly* from a random sample of 50 women in their average enjoyment of a new television show. Suppose that I arranged to have each sample view my new television show and than rate, on a scale from 1 to 10, how much they enjoyed the show, with higher scores indicating greater enjoyment. In addition, suppose that my sample of men gave the show an average rating of 7.5 and my sample of women gave the show an average rating of 6.5.

In looking at these two means, I can clearly see that my sample of men had a higher mean enjoyment of the television show than did my sample of women. But if you'll look closely at my earlier question, I did not ask simply whether my sample of men differed from my sample of women in their average enjoyment of the show. I asked whether they differed *significantly* in their average enjoyment of the show. The word *significantly* is critical in much of statistics, so I discuss it briefly here as it applies to independent *t* tests (for a more thorough discussion, see Chapter 6).

When I conduct an independent samples *t* test, I generally must collect data from two samples and compare the means of these two samples. But I am interested not only in whether these two samples differ on some variable. I am also interested in whether the differences in the two sample means are large enough to suggest that there are also differences in the two *populations* that these samples represent. So, returning to our previous example, I already know that the 50 men in my sample enjoyed the television show more, on average, than did the 50 women in my sample. So what? Who really cares about these 50 men and these 50 women, other than their friends and families? What I really want to know is whether the difference between these two samples of men and women is large enough to indicate that men *in general* (i.e., the population of men that this sample represents) will like the television show more than women *in general* (i.e., the population of women that this sample represents). In other words, is this difference of 1.0 between my two samples large enough to represent a real difference between the populations of men and women? The way of asking this question in statistical shorthand is to ask "Is the difference between the means of these two samples statistically significant?" (or **significant** for short).

To answer this question, I must know how much difference I should *expect* to see between two samples of this size drawn from these two populations? If I were to randomly select a different sample of 50 men and a different sample of 50 women, I might get the opposite effect, where the women outscore the men. Or, I might get an even larger difference, where men outscore the women by 3 points rather than 1. So the critical question here is this: What is the *average expected difference between the means* of two samples of this size (i.e., 50 each) selected randomly from these two populations? In other words, *what is the* **standard error of the difference between the means?**

As I have said before, understanding the concept of standard error provides the key to understanding how inferential statistics work, so take your time and reread the preceding four paragraphs to make sure you get the gist. Regarding the specific case of independent samples *t* tests, we can conclude that the question we want to answer is whether the difference between our two samples means is large or small compared to the amount of difference we would expect to see just by selecting two different samples. Phrased another way, we want to know whether our *observed* difference between our two samples means is large relative to the standard error of the difference between the means. The general formula for this question is as follows:

$$t = \frac{observed\ difference\ between\ sample\ means}{s\ tan\ dard\ error\ of\ the\ difference\ between\ the\ means}$$

or

$$t = \frac{\overline{X}_1 - \overline{X}_2}{s_{\overline{x}1-\overline{x}2}}$$

where $\overline{X}_1$ is the mean for sample 1

$\overline{X}_2$ is the mean for sample 2

$s_{\overline{x}1-\overline{x}2}$ is the standard error of the difference between the means

The Standard Error of the Difference Between Independent Sample Means

The standard error of the difference between independent samples means is a little bit more complex than the standard error of the mean discussed in Chapter 5. That's because instead of dealing with a single sample, now we have to find a single standard error involving two samples. Generally speaking, this involves simply combining standard errors of the two samples. In fact, when the two samples are roughly the same size, the standard error for the difference between the means really is similar to simply combining the two sample standard errors of the mean, as the formula presented in Table 8.1 indicates.

When the two samples are not roughly equal in size, there is a potential problem with using the formulas in Table 8.1 to calculate the standard error. Because these formulas essentially blend the standard errors of each sample together, they also essentially give each sample equal weight and treat the two samples as one new, larger sample. But if the two samples are not of equal size, and especially if they do not have equal standard deviations, than we must adjust the formula for the standard error to take these differences into account. The only difference between this formula and the formula for the standard error when the sample sizes are equal is that the unequal sample size formula adjusts for the different samples sizes.

In practice, let us hope that you will never need to actually calculate any of these standard errors by hand. Because computer statistical programs compute these for us these days, it may be more important to understand the concepts involved than the components of the formulas themselves. In this spirit, try to understand what the standard error of the difference between independent samples means is and why it may differ if the sample sizes are unequal. Simply put, the standard error of the difference between two independent samples means is the average expected difference between any two samples of a given size randomly selected from a population on a given variable. In our example comparing men's and women's enjoyment of the new television show, the standard error would be the average (i.e., *standard*) amount of difference (i.e., error) we would expect to find between any two samples of 50 men and 50 women selected randomly from the larger populations of men and women.

TABLE 8.1 Formula for Calculating the Standard Error of the Difference Between Independent Sample Means when the Samples Sizes Are Roughly Equal (i.e., $N_1 \approx N_2$).

$$s_{\bar{x}1-\bar{x}2} = \sqrt{s_{\bar{x}1}^2 + s_{\bar{x}2}^2}$$

$s_{\bar{x}1}$ is the standard error of the mean for the first sample

$s_{\bar{x}2}$ is the standard error of the mean for the second sample

Determining the Significance of the *t* Value for an Independent Samples *t* Test

Once we calculate the standard error and plug it into our formula for calculating the *t* value, we are left with an *observed t* value. How do we know if this *t* value is statistically significant? In other words, how do we decide if this *t* value is large enough to indicate that the difference between my sample means probably represents a real difference between my population means? To answer this question, we must find the probability of getting a *t* value of that size by chance. In other words, what are the odds that the difference between my two samples means is just due to the luck of the draw when I selected these two samples at random rather than some real difference between the two populations? Fortunately, statisticians have already calculated these odds for us, and a table with such odds is included in Appendix B. Even more fortunately, statistical software programs used on computers calculate these probabilities for us, so there will hopefully never be a need for you to use Appendix B. I provide it here so because I think the experience of calculating a *t* value by hand and determining whether it is statistically significant can help you understand how *t* tests work.

In Chapter 4, we saw how statisticians generated probabilities based on the normal distribution. With *t* distributions, the exact same principles are involved, except that now we have to take into account the size of the samples we are using. This is because the shape of the *t* distribution changes as the sample size changes, and when the shape of the distribution changes, so do the probabilities associated with it. The way that we take the sample size into account in statistics is to calculate degrees of freedom. The explanation of exactly what a degree of freedom is may be a bit more complicated than is worth discussing here (although you can read about it in most statistics textbooks if you are interested). At this point, suffice it to say that in an independent samples *t* test, you find the degrees of freedom by adding the two sample sizes together and subtracting 2. So the formula is $df = n_1 + n_2 - 2$. Once you have your degrees of freedom and your *t* value, you can look in the table of *t* values in Appendix B to see if the difference between your two sample means is significant.

To illustrate this, let's return to our example comparing men's and women's enjoyment of the new television program. Let's just suppose that the standard error of the difference between the means is .40. When I plug this number into the *t* value formula I get the following:

$$t = \frac{7.5 - 6.5}{.40}$$

$$t = \frac{1.0}{.40} = 2.50$$

$$df = 50 + 50 - 2 = 98$$

Now that we have a *t* value and our degrees of freedom, we can look in Appendix B to find the probability of getting a *t* value of this size ($t = 2.50$) by chance when we have 98 degrees of freedom. My table tells me that the probability of getting a *t* value this large by chance (i.e., due strictly to random sampling) is between 1% and 2%. In other words, when we randomly select two samples of 50 each from two different populations, we would expect to a *t* value of this size less than 2% of the time *when there is no real difference between the population means* (for a more thorough discussion of this issue, see Chapter 6). Because this is such a small probability, I conclude that the difference between my sample of 50 men and my sample of 50 women that I observed in the average ratings of enjoyment of the television show probably represents a real difference between the larger populations of men and women rather than some fluke difference that emerged simply because of who I happened to get in my samples (i.e., *random sampling error*).

It is important to remember that although this difference between the means was *statistically* significant, that does not necessarily mean that it is *practically* significant (refer to the discussion about effect size in Chapter 6). Just as the standard error of the mean is influenced by the size of the sample, the standard error of the *difference* between the means is also affected by sample size. The larger the samples, the smaller the standard error and the more likely it is that you will find a statistically significant result. To determine whether this difference between men and women is *practically* significant, we should consider the actual *raw score* difference. Men in our sample scored an average of one point higher on a 10-point scale than did women. Is that a big difference? Well, that is a judgment call. I would consider that a fairly inconsequential difference because we are talking about preferences for a television show. I don't consider a one-point difference on a 10-point scale regarding television preferences to be important. But potential advertisers might consider this a meaningful difference. Those wanting to advertise female-oriented products may not select this show, which seems to appeal more to male viewers.

Another way to determine whether this difference in the means is practically significant is to calculate an effect size. The formula for the effect size for an independent samples *t* test is presented in Table 8.2. To calculate the effect size, you must first calculate the denominator. Using our example where the sample size for one group is 50 and the standard error of the difference between the means is .40, we get the following:

$$\hat{s} = \sqrt{.50}(.40)$$

$$\hat{s} = 7.07(.40)$$

$$\hat{s} = 2.83$$

We can then plug this into the formula for the effect size, along with the two sample means:

$$d = \frac{7.5 - 6.5}{2.83} \Rightarrow d = .35$$

So our effect size for this problem is .35, which would be considered a small- to medium-size effect.

TABLE 8.2 Formula for the effect size for an independent samples *t* test.

$$d = \frac{\overline{X}_1 - \overline{X}_2}{\hat{s}}$$

$$\hat{s} = \sqrt{n_1}\,(s_{\overline{x}1-\overline{x}2})$$

where $\overline{X}_1$ is the mean for the first sample

$\overline{X}_2$ is the mean for the second sample

n_1 is the sample size *for one sample*

$(s_{\overline{x}1-\overline{x}2})$ is the standard error of the difference between the means

PAIRED OR DEPENDENT SAMPLES *t* TESTS IN DEPTH

Most of what I wrote before about the independent samples *t* test applies to the paired or dependent samples *t* test as well. We are still interested in determining whether the difference in the means that we observe in some sample(s) on some variable represents a true difference in the population(s) from which the sample(s) were selected. For example, suppose I wanted to know whether employees at my widget-making factory are more productive after they return from a 2-week vacation. I randomly select 30 of my employees and calculate the average number of widgets made by each employee during the week before they go on vacation. I find that, on average, my employees made 250 widgets each during the week. During the week after they return from vacation, I keep track of how many widgets is made by *the same sample of 30* employees and find that, on average, they made 300 widgets each during the week after returning from their vacations.

Just as with the independent samples *t* test, here I am concerned not only with whether this sample of 30 employees made more or fewer widgets after their vacation. I can look at the prevacation and postvacation averages and see that these 30 employees, on average, made an average of 50 more widgets a week after their vacation. That is quite a lot. But I also want to know whether what I observed in this sample represents a likely difference in the productivity of the larger population of widget makers after a vacation. In other words, is this a statistically significant difference? The only real distinction between this dependent samples *t* test and the independent samples *t* test is that rather than comparing two samples on a single dependent variable, now I am comparing the average scores of a single sample (i.e., the same group of 30 employees) on two variables (i.e., prevacation widget-making average and postvacation widget making average). To make this comparison, I will again need to conduct a *t* test in which I find the difference between the two means and divide by the standard error of the difference between two *dependent sample* means. This equation looks like this:

$$t = \frac{observed\ difference\ between\ pre-vacation\ and\ post-vacation\ means}{s\tan dard\ error\ of\ the\ difference\ between\ the\ means}$$

or

$$t = \frac{\overline{X} - \overline{Y}}{s_{\overline{D}}}$$

where $\overline{X}$ is the prevacation mean
$\overline{Y}$ is the postvacation mean
$s_{\overline{D}}$ is the standard error of the difference between the means

The formula for calculating the standard error of the difference between the means for dependent samples is slightly different than the one for independent samples, but the principles involved (i.e., what the standard error represents) are the same. Keep in mind that if I were to continually randomly select a sample of 30 widget makers and compare their prevacation and postvacation productivity, I could generate a distribution of difference scores. For some samples, there would be no difference between prevacation and postvacation productivity. For others, there would be increases in productivity and for still other samples there would be decreases in productivity. This distribution of difference scores (i.e., differences between prevacation and postvacation averages) would have a mean and a standard deviation. The standard deviation of this distribution would be the **standard error of the differences between dependent samples**. The formula for this standard error is presented below in Table 8.3.

TABLE 8.3   Formula for the standard error of the difference between
dependent sample means.

Step 1: $s_{\overline{D}} = \dfrac{s_D}{\sqrt{N}}$

Step 2: $s_D = \sqrt{\dfrac{\Sigma D^2 - \dfrac{(\Sigma D)^2}{N}}{N-1}}$

where $s_{\overline{D}}$ is the standard error of the difference between dependent sample means

s_D is the standard deviation of the difference between dependent sample means

D is the difference between each pair of X and Y scores (i.e., $X - Y$)

N is the number of pairs of scores

As you can see in Table 8.3, the easiest way to find the standard error is to follow a two-step process. First, we can find the standard deviation of difference scores for my sample. Then, we can divide this by the square root of the sample size to find the standard error. This formula is very similar to the formula for finding the standard error of the mean.

Another difference between dependent and independent samples t tests can be found in the calculation of the degrees of freedom. Whereas we had to add the two samples together and subtract 2 in the independent samples formula, for dependent samples we find the number of pairs of scores and subtract 1. In our example of widget makers, we have 30 pairs of scores because we have two scores for each person in the sample (one prevacation score and one postvacation score). In the case of a paired t test where we have two paired samples (e.g., fathers and their sons), we use the same formula for calculating the standard error and the degrees of freedom. We must simply remember to match each score in one sample with a corresponding score in the second sample (e.g., comparing each father's score with only his son's score).

Once we've found our t value and degrees of freedom, the process for determining the probability of finding a t value of a given size with a given number of degrees of freedom is exactly the same as it was for the independent samples t test.

EXAMPLE: COMPARING BOYS' AND GIRLS' GRADE POINT AVERAGES

To illustrate how t tests work in practice, I provide one example of an independent samples t test and one of a dependent samples t test using data from a longitudinal study conducted by Carol Midgley and her colleagues. In this study, a sample of students were given surveys each year for several years beginning when the students were in the fifth grade. In the examples that follow, I present two comparisons of students' GPAs. The GPA is an average of students' grades in the four core academic areas: math, science, English, and social studies. Grades were measured using a 13-point scale with 13 = "A+" and 0 = "F".

In the first analysis, an independent samples t test was conducted to compare the average grades of 6th grade boys and girls. This analysis was conducted using SPSS computer software. Thankfully, this program computes the means, standard error, t value, and probability of obtaining the t value by chance. Because the computer does all of this work, there is nothing to compute by hand, and I can focus all of my energy on interpreting the results. I present the actual results from the t test conducted with SPSS in Table 8.4.

TABLE 8.4   SPSS results of independent samples *t* test.

Variable	Number of Cases	Mean	SD	SE of Mean
Sixth Grade GPA				
Male	361	6.5783	2.837	.149
Female	349	8.1387	2.744	.147

Mean Difference = -1.5604

Levene's Test for Equality of Variances: *F* = .639   *p* = .424

t test for Equality of Means

Variances	t Value	df	2-Tail Sig	SE of Diff
Equal	-7.45	708	.000	.210
Unequal	-7.45	708.00	.000	.209

SPSS presents the sample sizes for boys (*n* = 361) and girls (*n* = 349) first, followed by the mean, standard deviation ("SD"), and standard error of the mean ("SE of mean") for each group. Next, SPSS reports the actual difference between the two sample means ("Mean Difference = -1.5604"). This mean difference is negative because boys are the X_1 group and girls are the X_2 group. Because girls have the higher mean, when we subtract the girls mean from the boys mean (i.e., $\overline{X}_1 - \overline{X}_2$) we get a negative number. Below the mean difference we see the "Levene's Test for Equality of Variances." [5] This test tells us that there is not a significant difference between the standard deviations of the two groups on the dependent variable (GPA). Below the test for equality of variances SPSS prints two lines with the actual *t* value (-7.45), the degrees of freedom ("*df*" = 708), the *p* value ("2-Tail Sig" = .000), and the standard error of the difference between the means ("SE of Diff" = .210 and .209). These two lines of statistics are presented separately depending on whether we have equal or unequal variances. Because we had equal variances (as determined by Levene's test), we should interpret the top line, which is identified by the "Equal" name in the left column. Notice that these two lines of statistics are almost identical. That is because the variances are not significantly different between the two groups. If they had been different, the statistics presented in these two lines would have differed more dramatically.

If we take the difference between the means and divide by the standard error of the difference between the independent sample means, we get the following equation for *t*:

$$t = -1.5604 \div .210$$

$$t = -7.45$$

---

[5] One final assumption that must be considered when conducting independent samples *t* tests is that the variances, or standard deviations, of the dependent variable are equal between the two samples. In other words, it is important to know whether the scores on the dependent variable are more varied in one sample than in the other. That is because when we calculate the standard error for the independent samples *t* test, we are basically combining the standard errors from the two samples. Because standard errors are determined in part by the size of the standard deviation, if the standard deviations for the two samples are very different, when we combine them the samples will not provide as accurate an estimate of the population as they would have had they been similar. To adjust for this, we must reduce our degrees of freedom when the variances of our two samples are not equal. SPSS does this automatically, as the example presented in Table 8.4 indicates.

The probability of getting a *t* value of –7.45 with 708 degrees of freedom is very small, as our *p* value ("2-Tail Sig") of .000 reveals. Because *t* distributions are symmetrical (as are normal distributions), there is the exact same probability of obtaining a given negative *t* value by chance as there is of obtaining the same positive *t* value. For our purposes, then, we can treat negative *t* values as absolute numbers.[6]

The results of the *t* test presented in Table 8.4 indicate that our sample of girls had higher average GPAs than did our sample of boys, and that this difference was statistically significant. In other words, if we kept randomly selecting samples of these sizes from the larger populations of sixth grade boys and girls and comparing their average GPAs, the odds of finding a difference between the means that is this large *if there is no real difference between the means of the two populations* is .000. This does not mean there is absolutely no chance. It just means that SPSS does not print probabilities smaller than .001 (e.g., .00001). Because this is such a small probability, we conclude that the difference between the two sample means probably represents a genuine difference between the larger populations of boys and girls that these samples represent. Girls have *significantly* higher GPA's than boys. Reminder: Statistical significance is influenced by sample size. Our sample size was quite large, so a difference of about 1.56 points on a 14-point scale was statistically significant. But is it practically significant? You can compute an effect size to help you decide.

EXAMPLE: COMPARING FIFTH AND SIXTH GRADE GPA

Our second example involves a comparison of students' grade point averages in fifth grade with the same sample's GPAs a year later, at the end of sixth grade. For each student in the sample (*N* = 690), there are two scores: one GPA for fifth grade, one GPA for sixth grade. This provides a total of 689 pairs of scores, and leaves us with 688 degrees of freedom (*df* = number of pairs – 1). A quick glance at the means reveals that, in this sample, students had slightly higher average GPA's in fifth grade (8.0800) than they did a year later in sixth grade (7.3487). But is this a *statistically significant* difference? To know, we must conduct a dependent samples *t test*, which I did using SPSS (see Table 8.5).

TABLE 8.5 SPSS results for dependent samples *t* test.

Variable	Number of Pairs	Corr	2-Tail Sig.	Mean	SD	SE of Mean
GPA5.2				8.0800	2.509	.096
	689	.635	.000			
GPA6.2				7.3487	2.911	.111

Paired Differences						
Mean	SD	SE of Mean		t Value	df	2-Tail Sig.
.7312	2.343	.089		8.19	688	.000

This analysis produced a *t* value of 8.19, which my SPSS program told me had a probability of occurring less than one time in a thousand due to chance ("2-tail Sig" = .000). Therefore, I

---

[6] Sometimes researchers must concern themselves with whether the t-value is positive or negative. In statistics, there are "two-tailed" hypotheses in which the researcher simply wants to know if two groups differ, without speculating about which group may be higher. When there are two-tailed hypotheses, it does not matter whether the t-value is positive or negative. Other times, however, the researcher may have a "one-tailed" hypotheses in which a direction of difference is postulated. For example, I could have tested the hypotheses that boys have higher GPA's than girls rather than simply examined whether the two groups differed somehow. When testing a one-tailed hypothesis, it is essential that one consider whether the t-value is positive or negative. For a more thorough discussion of one-tailed and two-tailed hypotheses, consult any traditional statistics textbook, such as *Statistical methods for the social and behavioral sciences* by Marascuilo and Serlin (1988).

conclude that the difference between fifth and sixth grade GPA's in my sample probably represents a real difference between the GPA's of the larger population of fifth and sixth graders that my sample represents. Although this difference is statistically significant, notice that it is a difference of only about .73 points on a 14-point scale. Also notice that the SPSS program also provides a measure of the correlation between the two variables ("corr" = .635) and indicates that this correlation coefficient is statistically significant. This tells you that students' 5th grade GPA is strongly related to their 6th grade GPA, as you might expect. Finally, notice that at the bottom left of Table 8.5, the difference between the means ("Paired Differences Mean"), the standard deviation of the difference between the means ("SD"), and the standard error of the difference between the means ("SE of Mean") are all presented. The differences between the means divided by the standard error of the difference between the means produces the *t* value.

WRAPPING UP AND LOOKING FORWARD

The two different types of *t* tests described in this chapter share two things in common. First, they both test the equality of means. Second, they both rely on the *t* distribution to produce the probabilities used to test statistical significance. Beyond that, these two different types of *t* tests are really quite different. One, the independent samples *t* test, is used to examine the equality of means from two independent groups. Such a test has much in common with one-way ANOVA (Chapter 9) and factorial ANOVA (Chapter 10). In contrast, the dependent samples *t* test is used to examine whether the means of *related* groups, or of two variables examined within the same group, are equal. This test is more directly related to repeated-measures ANOVA as discussed in Chapter 11.

GLOSSARY OF TERMS AND SYMBOLS FOR CHAPTER 8

Categorical, nominal: When variables are measured using categories, or names.

Continuous, intervally scaled: When variables are measured using numbers along a continuum with equal distances, or values, between each number along the continuum.

Dependent variable: A variable for which the values may depend on, or differ by, the value of the independent variable. When the dependent variable is statistically related to the independent variable, the value of the dependent variable "depends" on, or is predicted by, the value of the independent variable.

Independent samples *t* test: A test of the statistical similarity between the means of two independent samples on a single variable.

Independent variable: A variable that may predict or produce variation in the dependent variable. The independent variable may be nominal or continuous and is sometimes manipulated by the researcher (e.g., when the researcher assigns participants to an experimental or control group, thereby creating a two-category independent variable).

Matched, paired, dependent samples: When each score of one sample is matched to one score from a second sample. Or, in the case of a single sample measured at two times, when each score at Time 1 is matched with the score for the same individual at Time 2.

Matched, paired, dependent samples *t* test: Test comparing the means of paired, matched, or dependent samples on a single variable.

Significant: Shortened form of the expression "statistically significant."

Standard error of the difference between the means: A statistic indicating the standard deviation of the sampling distribution of the difference between the means.

$s_{\bar{x}_1 - \bar{x}_2}$ Symbol for the standard error of difference between two independent sample means.

$s_{\bar{D}}$ Symbol for the standard error of the difference between two dependent, matched, or paired samples.

s_D Symbol for the standard deviation of the difference between two dependent, matched, or paired samples.

df   Symbol for degrees of freedom.

t   Symbol for the *t* value.

CHAPTER 9

ONE-WAY ANALYSIS OF VARIANCE

The purpose of a one-way analysis of variance (**one-way ANOVA**) is to compare the means of two or more groups (the independent variable) on one dependent variable to see if the group means are significantly different from each other. In fact, if you want to compare the means of two independent groups on a single variable, you can use either an independent samples t test or a one-way ANOVA. The results will be identical, except instead of producing a t value, the ANOVA will produce an F ratio, which is simply the t value squared (more about this in the next section of this chapter). Because the t test and the one-way ANOVA produce identical results when there are only two groups being compared, most researchers use the one-way ANOVA only when they are comparing three or more groups. To conduct a one-way ANOVA, you need to have a **categorical (or nominal) variable** that has at least two independent groups (e.g., a race variable with the categories African-American, Latino, and Euro-American) as the independent variable and a continuous variable (e.g., achievement test scores) as the dependent variable.

Because the independent t test and the one-way ANOVA are so similar, people often wonder, Why don't we just use t tests instead of one-way ANOVAs? Perhaps the best way to answer this question is by using an example. Suppose that I want to go into the potato chip business. I've got three different recipes, but because I'm new to the business and don't have a lot of money, I can produce only one flavor. I want to see which flavor people like best and produce that one. I randomly select 90 adults and randomly divide them into three groups. One group tries my BBQ-flavored chips, the second group tries my ranch-flavored chips, and the third group tastes my cheese-flavored chips. All participants in each group fill out a rating form after tasting the chips to indicate how much they liked the taste of the chips. The rating scale goes from a score of 1 ("Hated it") to 7 ("Loved it"). I then compare the average ratings of the three groups to see which group liked the taste of their chips the most. In this example, the chip flavor (BBQ, Ranch, Cheese) is my categorical, independent variable and the rating of the taste of the chips is my continuous, dependent variable.

To see which flavor received the highest average rating, I could run three separate independent t tests comparing (a) BBQ with Ranch, (b) BBQ with Cheese, and (c) Ranch with Cheese. The problem with running three separate t tests is that each time we run a t test, we must make a decision about whether the difference between the two means is meaningful, or statistically significant. This decision is based on probability, and every time we make such a decision, there is a slight chance we might be wrong (see Chapter 6 on statistical significance). The more times we make decisions about the significance of t tests, the greater the chances are that we will be wrong. In other words, the more t tests we run, the greater the chances become of deciding that a t test is significant (i.e., that the means being compared are really different) when it really is not. A one-way ANOVA fixes this problem by adjusting for the number of groups being compared. To see how it does this, let's take a look at one-way ANOVA in more detail.

ONE-WAY ANOVA IN DEPTH

The purpose of a one-way ANOVA is to divide up the variance in some dependent variable into two components: the variance attributable to **between-group** differences, and the variance attributable to **within-group** differences, also known as *error*. When we select a sample from a population and calculate the mean for that sample on some variable, that sample mean is our best predictor of the population mean. In other words, if we do not know the mean of the population, our best guess about what the population mean is would have to come from the mean of a sample drawn randomly from that population. Any scores in the sample that differ from the sample mean are believed to include what statisticians call error. For example, suppose I have a sample of 20 randomly selected fifth graders. I give them a test of basic skills in math and find out that, in my sample, the average number of items answered correctly on my test is 12. If I were to select one student in my sample and find that she had a score of 10 on the test, the difference between her score and the sample mean would be considered error.

The variation that we find among the scores in a sample is not just considered error. In fact, it is thought to represent a specific kind of error: *random error*. When we select a sample at random from a population, we expect that the members of that sample will not all have identical scores on our variable of interest (e.g., test scores). That is, we expect that there will be some variability in the scores of the members of the sample. That's just what happens when you select members of a sample randomly from a population. Therefore, the variation in scores that we see among the members of our sample is just considered random error.

The question that we can address using ANOVA is this: Is the average amount of difference, or variation, between the scores of members of *different* samples large or small compared to the average amount of variation *within* each sample, otherwise known as random error (a.k.a. error)? To answer this question, we have to determine three things. First, we have to calculate the average amount of variation within each of our samples. This is called the **mean square within** (MS_w) or the **mean square error** (MS_e). Second, we have to find the average amount of variation *between* the groups. This is called the **mean square between** (MS_b). Once we've found these two statistics, we must find their ratio by dividing the mean square between by the mean square error. This ratio provides our **F value**, and when we have our *F* value we can look at our family of *F* distributions to see if the differences between the groups are statistically significant (see Table 9.1).

TABLE 9.1 Formula for the *F* value.

$$F = \frac{mean\ square\ between}{mean\ square\ error}$$

or

$$F = \frac{MS_b}{MS_e}$$

where *F* is the *F* value
MS_b is mean square between group
MS_e is mean square error, or within groups

Note that, although it may sound like analysis of variance is a whole new concept, in fact it is virtually identical to the independent *t* test discussed in Chapter 8. Recall that the formula for calculating an independent *t* test also involves finding a ratio. The top portion of the fraction is the difference between two sample means, which is analogous to the mean square between (MS_b) just presented. The only differences between the two are (a) rather than finding a simple difference between two means as in a *t* test, in ANOVA we are finding the *average* difference between means, because we often are comparing more than two means; and (b) we are using the squared value of the difference between the means. The bottom portion of the fraction for the *t* test is the standard *error* of the difference between two sample means. This is exactly the same as the *average*, or standard, error within groups. In the formula used to calculate the *F* value in ANOVA, we must square this average within-group error. So, just as in the *t* test, in ANOVA we are trying to find the average difference *between* group means relative to the average amount of variation *within* each group.

To find the MS_e and MS_b, we must begin by finding the **sum of squares error** (SS_e) and the **sum of squares between** (SS_b). This sum of squares idea is not new. It is the same old sum of squares introduced back in Chapter 2 when talking about variance and standard deviation. Sum of squares is actually short for *sum of squared deviations.* In the case of ANOVA, we have two types of deviations. The first is deviations between each score in a sample and the mean for that sample (i.e., error). The second type of deviation is between each sample mean and the mean for all of the groups combined, called the **grand mean** (i.e., between groups). To find the sum of squares error (SS_e):

1. Subtract the group mean from each individual score in each group: $(X - \overline{X})$.
2. Square each of these deviation scores: $(X - \overline{X})^2$.
3. Add them all up for each group: $\Sigma(X - \overline{X})^2$.
4. Then add up all of the sums of squares for all of the groups:
   $\Sigma(X_1 - \overline{X}_1)^2 + \Sigma(X_2 - \overline{X}_2)^2 + \ldots + \Sigma(X_k - \overline{X}_k)^2$

Note: The subscripts indicate the individual groups, through the last group, which is indicated with the subscript k.

The method used to calculate the sum of squares between groups (SS_b) is just slightly more complicated than the SS_e formula. To find the SS_b we

1. Subtract the grand mean from the group mean: $(\overline{X} - \overline{X}_T)$; $_T$ indicates total, or the mean for the total group.
2. Square each of these deviation scores: $(\overline{X} - \overline{X}_T)^2$.
3. Multiply each squared deviation by the number of cases in the group: $[n(\overline{X} - \overline{X}_T)^2]$.
4. Add these squared deviations from each group together: $\Sigma[n(\overline{X} - \overline{X}_T)^2]$.

The only real differences between the formula for calculating the SS_e and the SS_b are:

1. In the SS_e we subtract the group mean from the individual scores in each group whereas in the SS_b we subtract the grand mean from each group mean.
2. In the SS_b we multiply each squared deviation by the number of cases in each group. We must do this to get an approximate deviation between the group mean and the grand mean *for each case in every group.*

If we were to add the SS_e to the SS_b, the resulting sum would be called the **sum of squares total (SS_T)**. A brief word about the SS_T is in order. Suppose that we have three randomly selected samples of children. One is a sample of 5th graders, another is a sample of 8th graders, and the third is a sample of 11th graders. If we were to give each student in each sample a spelling test, we could add up the scores for all of the children in the three samples combined and divide by the total number of scores to produce one average score. Because we have combined the scores from all three samples, this overall average score would be called the grand mean, or total mean, which would have the symbol $\overline{X}_T$. Using this grand mean, we could calculate a squared deviation score for each child in all three of our samples combined using the familiar formula $(X - \overline{X}_T)^2$. The interesting thing about these squared deviations is that, for each child, the difference between each child's score and the grand mean is the sum of that child's deviation from the mean of his or her own group plus the deviation of that group mean from the grand mean. So, suppose Jimmy is in the fifth-grade sample. Jimmy gets a score of 25 on the spelling test. The average score for the fifth-grade sample is 30, and the average score for all of the samples combined (i.e., the grand mean) is 35. The difference between Jimmy's score (25) and the grand mean (35) is just the difference between Jimmy's score and the mean for his group (25 – 30 = -5) plus the difference between his group's mean and the grand mean (30 – 35 = -5). Jimmy's deviation from the grand mean is –10. If we square that deviation score, we end up with a squared deviation of 100 for Jimmy.

Now, if we calculated a deviation score for each child in all three samples and added up all of these deviation scores using the old $\Sigma(X - \overline{X}_T)^2$ formula, the result would be the sum of squares total, or the SS_T. (Notice that this formula is the same one that we used way back in Chapter 2! It is the numerator for the variance formula!) The interesting thing about this SS_T is that it is really just the sum of the SS_b and the SS_e. $SS_T = SS_b + SS_e$. This makes sense, because, as we saw with Jimmy, the difference between any individual score and the grand mean is just the sum of the difference between the individual score and the mean of the group that the individual is from plus the difference between that group mean and the grand mean. This is the crux of ANOVA.

Deciding If the Group Means Are Significantly Different

Once we have calculated the SS_b and the SS_e, we have to convert them to average squared deviation scores, or MS_b and MS_e. This is necessary because there are far more deviation scores in the SS_e than there are in the SS_b, so the sums of squares can be a bit misleading. What we want to know in an ANOVA is whether the *average* difference between the group means is large or small relative to the *average* difference between the individual scores and their respective group means, or the average amount of error within each group. To convert these sums of squares into mean squares, we must divide the sums of squares by their appropriate degrees of freedom.

For the SS_b, remember that we are only making comparisons between each of the groups. The degrees of freedom for the SS_b is always the number of groups minus 1. If we use K to represent the number of groups, and df to represent degrees of freedom, then the formula for the between groups degrees of freedom is $df = K - 1$. So, to convert an SS_b to an MS_b, we divide SS_b by $K - 1$. The degrees of freedom for SS_e is found by taking the number of scores in each group and subtracting 1 from each group. So, if we have three groups, our df for SS_e will be $(n_1 - 1) + (n_2 - 1) + (n_3 - 1)$. Notice that this is the same formula for the degrees of freedom that was used for the independent samples t test in Chapter 8. The only difference is that we have one more group here. A simpler way to write this df formula is $N-K$, where N is the total number of cases for *all* groups combined and K is the number of groups. Once we have this df, we can convert the SS_e into an MS_e by simply dividing SS_e by $N - K$. Table 9.2 contains a summary of the formulas for converting the sums of squares into mean squares.

TABLE 9.2   Converting sums of squares into mean squares.

$$MS_b = \frac{SS_b}{K-1}$$	$$MS_e = \frac{SS_e}{N-K}$$
MS_b = Mean squares between groups SS_b = Sum of squares between groups N = The number of cases combined across all groups	MS_e = Mean squares error SS_e = Sum of squares error K = The number of groups N = The number of cases combined across all groups

Once we have found our MS_b and our MS_e, all we have to do is divide MS_b by MS_e to find our F value. Once we've found our F value, we need to look in our table of F values (Appendix C) to see whether it is statistically significant. This table of F values is similar to the table of t values we used in Chapter 8, with one important difference. Unlike t values, the significance of F values depends on both the number of cases in the samples (i.e., the df for MS_e) *and* the number of groups being compared (i.e., the df for MS_b). This second df is critical, because it is what is used to control for the fact that we are comparing more than two groups. Without it, we might as well conduct multiple t tests, and this is problematic for the reasons discussed at the beginning of the chapter. In Appendix C, we can find critical values for F associated with different alpha levels. If our observed value of F is larger than our critical value of F, we must conclude that there are statistically significant differences between the group means.

Post-Hoc Tests

Our work is not done once we have found a statistically significant difference between the group means. Remember that when we calculated MS_b, we ended up with an <u>average</u> difference between the group means. If we are comparing three group means, we might find a relatively large average difference between these group means even if two of the three group means are identical. Therefore, a statistically significant F value tells us only that somewhere there is a meaningful difference between my group means. But it does not tell us *which* groups differ from each other significantly. To do this, we must conduct **post-hoc tests**.

There are a variety of post-hoc tests available. Some are more conservative, making it more difficult to find statistically significant differences between groups, whereas others are more liberal. All post-hoc tests use the same basic principle. They allow you to compare each group mean to each other group mean and determine if they are significantly different while controlling for the number of group comparisons being made. As we saw in Chapters 6 and 8, to determine if the difference between two group means is statistically significant, we subtract one group mean from the other and divide by a standard error. The difference between the various types of post-hoc tests is what each test uses for the standard error. You should consult a traditional textbook for a discussion on the various types of post-hoc tests that are used. In this book, for the purposes of demonstration, we will consider the **Tukey HSD** (HSD stands for Honestly Significantly Different) post-hoc test.

The Tukey test compares each group mean to each other group mean by using the familiar formula described for t tests in Chapter 8. Specifically, it is the mean of one group minus the mean of a second group divided by the standard error:

$$Tukey\ HSD = \frac{\overline{X}_1 - \overline{X}_2}{s_{\overline{x}}}$$

$$\text{where } s_{\overline{x}} = \frac{MS_e}{N_T}$$

and N_T = the number of cases *in each group*

When we solve this equation, we get an *observed* Tukey HSD value. To see if this observed value is significant, and therefore indicating a statistically significant difference between the two groups being compared, we must compare our observed Tukey HSD value with a critical value. We find this critical value in Appendix D, which we read in pretty much the same way that we read the F-value table. That is, the number of groups being compared is listed on the top row of the table and the df_e is along the left column. In this table, only the critical values for an alpha level of .05 are presented.

Once we have calculated a Tukey HSD for each of the group comparisons we need to make, we can say which groups are significantly different from each other on our dependent variable. Notice that, because the standard error used in the Tukey HSD test assumes that each group has an equal number of cases, this is not the best post-hoc test to use if you have groups with unequal sample sizes.

A Priori Contrasts

Post-hoc tests such as the Tukey HSD automatically compare each group in the study with each other group. Sometimes, however, researchers are interested in knowing whether particular groups, or combinations of groups, differ from each other in their averages on the dependent variable. These analyses are known as *a priori* **contrasts**. Although such comparisons are generally conducted *after* the overall ANOVA has been conducted, they are called *a priori* contrasts because they are guided by research questions and hypothesis that were stated before the analyses were conducted. For example, suppose that I wanted to know whether children in different cities differed in their love for pepperoni pizza. I collect random samples of 10-year-olds from San Francisco, Chicago, Paris, and Rome. I ask all of the children to rate how much they like pepperoni pizza on a scale from 1 ("Hate it") to 20 ("Love it"). Because American children tend to eat a lot of junk food, I hypothesize that

American children, regardless of which American city they come from, will like pepperoni pizza more than European children. To test this hypothesis, I would contrast the average ratings of my American samples, combined across the two American cities, with those of the European children combined across the two European cities. Alternatively, I might hypothesize that children from Rome will report liking pepperoni pizza more than children in the other three cities, on average, because Italy is the birthplace of pizza. To test this, I could contrast the mean of my sample of children from Rome with the mean of my samples from the other three cities combined. Such contrasts allow researchers to test specific hypotheses regarding differences between the groups in their studies.

Effect Size

In addition to the calculation of effect size (d) presented in Chapters 6 and 8, another common measure of effect size is the percentage of variance in the dependent variable that is explained by the independent variable(s). To illustrate how this works, I present the results of an analysis using the SPSS computer software program to analyze a set of fictional data that I made up.

Suppose that I wanted to test a drug that I developed to increase students' interest in their schoolwork. Imagine that I randomly select 75 third-grade students and randomly assign them to one of three groups. The first group gets a relatively high dosage of the drug and is therefore called the "High Dose" group. The second group gets a low dosage of the drug and is called the "Low Dose" group. And the third group receives a placebo and is called the "Placebo" group. After dividing students into their respective groups, I give them the appropriate dosage of my new drug (or a placebo) and then give them all the exact same schoolwork assignment. I measure their interest in the schoolwork by asking them to rate how interesting they thought the work was on a scale from 1 ("Not interesting") to 5 ("Very interesting"). Then I use SPSS to conduct an ANOVA on my data and I get the output from the program presented in Table 9.3.

TABLE 9.3 SPSS output for ANOVA examining interest by drug treatment group.

Descriptive Statistics

Independent Variable	Mean	Std. Deviation	N		
High Dose	2.7600	1.2675	25		
Low Dose	3.6000	1.2583	25		
Placebo	2.6000	.9129	25		
Total	2.9867	1.2247	75		

ANOVA Results

Source	Type III Sum of Squares	df	Mean Square	F	Sig.	Eta Squared
Corrected Model	14.427	2	7.213	5.379	.007	.130
Intercept	669.013	1	669.013	498.850	.000	.874
Group	14.427	2	7.213	5.379	.007	.130
Error	96.560	72	1.341			

The results produced by SPSS include descriptive statistics such as the means, standard deviations, and sample sizes for each of the three groups as well as the overall mean ("Total") for the entire sample of 75 students. In the descriptive statistics we can see that the "Low Dose" group has a somewhat higher average mean on the dependent variable (i.e., interest in the schoolwork) than do the other two groups. Turning now to the ANOVA results below the descriptive statistics in Table 9.3, there are identical statistics for the "Corrected Model" row and the "Group" row. The "Model" row includes all effects in the model, such as all independent variables and interaction effects (see Chapter 10 for a discussion of these multiple effects). In the present example, there is only one independent variable, so the "Model" statistics are the same as the "Group" statistics.

Let's focus on the "Group" row. This row includes all of the between-group information, because "Group" is our independent group variable. Here we see the Sum of Squares[7] between (SS_b), which is 14.427. The degrees of freedom ("df") here is 2, because with three groups, $K - 1 =$ 2. The sum of squares divided by degrees of freedom produces the mean square (MS_b), which is 7.213. The statistics for the sum of squares error (SS_e), degrees of freedom for the error component, and mean square error (MS_e) are all in the row below the "Group" row. The F value ("F") for this ANOVA is 5.379, which was produced by dividing the mean square from the "Group" row by the mean square from the error row. This F value is statistically significant ("Sig." = .007). Finally, in the "Eta Squared" column, we can see that we have a value of .130 in the "Group" row. Eta squared is a measure of the association between the independent variable ("Group") and the dependent variable ("Interest"). It indicates that 13% of the variance in the scores on the interest variable can be explained by the Group variable. In other words, I can account for 13% of the variance in the interest scores simply by knowing whether students were in the "High Dose," "Low Dose," or "Placebo" group. Eta squared is essentially the same as the coefficient of determination (r^2) discussed in Chapter 7 and again in Chapter 12.

TABLE 9.4 SPSS results of Tukey HSD post-hoc tests.

(I) Treatment 1, Treatment 2, Control	(J) Treatment 1, Treatment 2, Control	Mean Difference (I - J)	Std. Error	Sig.
High Dose	Low Dose	-.8400	.328	.033
	Placebo	.1600	.328	.877
Low Dose	High Dose	.8400	.328	.033
	Placebo	1.0000	.328	.009
Placebo	High Dose	-.1600	.328	.877
	Low Dose	-1.0000	.328	.009

Now that we know that there is a statistically significant difference between the three groups in their level of interest, and that group membership accounts for 13% of the variance in interest scores, it is time to look at our Tukey post-hoc analysis to determine which groups significantly differ from each other. The SPSS results of this analysis are presented in Table 9.4. The far left column of this table contains the reference group, and the column to the right of this shows the comparison groups. So in the first comparison, the mean for the "High Dose" group is compared to the mean for the "Low Dose" group. We can see that the "Mean Difference" between these two groups is -.8400, indicating that the "High Dose" group had a mean that was .84 points

---

[7] SPSS generally reports Type III sum of squares. This sum of squares is known as the "residual" sum of squares because it is calculated after taking the effects of other independent variables, covariates, and interaction effects into account.

lower than the mean of the "Low Dose" group on the interest variable. In the last column, we can see that this difference is statistically significant ("Sig." = .033). So we can conclude that students in the "Low Dose" group, on average, were more interested in their work than were students in the "High Dose" group. In the next comparison between "High Dose" and "Placebo" we find a mean difference of .16, which was not significant ("Sig." = .877). Looking at the next set of comparisons, we see that the "Low Dose" group is significantly different from both the "High Dose" group (we already knew this) and the "Placebo" group. At this point, all of our comparisons have been made and we can conclude that, on average, students in the "Low Dose" group were significantly more interested in their work than were students in the "High Dose" and "Placebo" groups, but there was no significant difference between the interest of students in the "High Dose" and "Placebo" groups.

EXAMPLE: COMPARING THE PREFERENCES OF 5-, 8-, AND 12-YEAR-OLDS

Suppose that I've got three groups: 5-year-olds, 8-year-olds, and 12-year-olds. I want to compare these groups in their liking of bubble gum ice cream, on a scale from 1 to 5. I get the data presented in Table 9.5. These data provide all of the pieces needed to calculate the F ratio, which is also presented in the bottom right corner of the table. Notice that from the individual scores presented for each group, all of the additional data can be calculated. To find the grand mean, we add all 15 scores in the table together and divide by 15, which is the number of scores. It is also a simple matter to find the means for each of the three groups. Once we have these means, we can begin calculating deviation scores, first for the error and then for the between groups (see the last two rows, column one in the table for examples of these calculations).

TABLE 9.5   Data for 5-, 8-, and 12-year-olds liking of bubble gum ice cream.

5-Year-Olds	8-Year-Olds	12-Year-Olds	Grand Mean	Mean Squares
5	5	4	3.466	$MS_e = 10.8 \div 12 = .90$
5	4	3		
4	4	2		$MS_b = 8.93 \div 2 = 4.47$
4	3	2		
3	3	1		
$Mean_1 = 4.2$	$Mean_2 = 3.8$	$Mean_3 = 2.4$		
$SS_{e1} = 2.8$	$SS_{e2} = 2.8$	$SS_{e3} = 5.2$	$SS_e = 10.8$	$SS_T = 19.73$
$.64 + .64 + .04 +$			$2.8 + 2.8 + 5.2$	$10.8 + 8.93$
$04 + 1.44 = 2.8$			$= 10.8$	
$SS_{b1} = 2.68$	$SS_{b2} = .55$	$SS_{b3} = 5.7$	$SS_b = 8.93$	$F = 4.47 \div .90 = 4.97$
$5(.73^2) =$	$5(.11) = .55$	$5(1.14) = 5.7$	$2.68 + .55 + 5.7$	
$5(.54) = 2.68$			$= 8.93$	

Once we've calculated the SS_e and the SS_b, we need to convert these scores to mean squares values. In the right-hand column of the table, these calculations are performed. To find the MS_e, we divide the SS_e by the degrees of freedom (N–K). With 15 cases ($N = 15$), and 3 groups ($K = 3$), we have 12 degrees of freedom for our error (15-3 = 12). When we divide our SS_e by our degrees of freedom, we get $10.8 \div 12 = .90$. So our MS_e is .90. To find our MS_b we need to divide our SS_b by our degrees of freedom between groups, which is $K - 1$. With three groups, our $K = 3$, and $K - 1 = 2$. From the table we find that our $SS_b = 8.93$, so our MS_b must equal $8.93 \div 2 = 4.47$.

Once we have our MS_b and our MS_e, we simply divide one by the other to find our F value. In this example, $F = 4.47 \div .90 = 4.97$. Now that we've found our F value, we need to determine whether it is statistically significant. To do this, we look in our table of F distribution values presented in Appendix C. Our F value has 2 degrees of freedom, one for the numerator and one for the denominator. Our numerator in the F ratio is the MS_b, and this has an associated degrees of freedom of 2 (because $K - 1 = 2$ when we have three groups). Our denominator is MS_e, which has an associated degrees of freedom of 12 (because $N - K = 12$ when we have 15 cases and 3 groups). So our F value has 2 and 12 degrees of freedom. We would write this statistic as follows: $F_{(2, 12)} = 4.97$. When we look at the values in Appendix C, we can see that this F value with 2 and 12 degrees of freedom would fall between the .05 and .01 alpha levels. So this F ratio would be statistically significant if our alpha (α) level was .05, but not if we selected a .01 alpha level.

Assuming we selected an alpha level of .05, we now know that we have a statistically significant F value. This tells us that there is a statistically significant difference between the means of our three groups in their liking of bubble gum ice cream. But I do not yet know which of my three groups differ from each other. To figure this out, I need to conduct post-hoc tests. So, I conduct Tukey tests to compare my three groups to each other. Recall that the formula for the Tukey test is the mean of one group minus the mean of another group divided by the standard error. When all of our groups have equal numbers of cases, then the standard error for the Tukey test is the same for each comparison of groups. In our example, we have equal numbers of cases in each group, so we only need to calculate the standard error once:

$$s_{\bar{x}} = \sqrt{\frac{MS_e}{N_T}}$$

$$s_{\bar{x}} = \sqrt{\frac{.90}{5}}$$

$$s_{\bar{x}} = \sqrt{.18} \implies s_{\bar{x}} = .42$$

With our standard error for the Tukey tests in place, we can compare the means for each of the three groups.

$$\text{Tukey}_{1-2} = \frac{4.2 - 3.8}{.42} \implies \frac{.4}{.42} \implies .95$$

$$\text{Tukey}_{1-3} = \frac{4.2 - 2.4}{.42} \implies \frac{1.8}{.42} \implies 4.29$$

$$\text{Tukey}_{2-3} = \frac{3.8 - 2.4}{.42} \implies \frac{1.4}{.42} \implies 3.33$$

The final step in our analysis is to determine whether each of these Tukey HSD values is statistically significant. To do this, we must look at the table of critical values for the **studentized range statistic** in Appendix D. The values in this table are organized in a similar way to those presented in the table of F values in Appendix C. However, instead of using the degrees of freedom between groups to find the appropriate column, we use the number of groups. In this example, we have three groups, so we find the column labeled "3." To find the appropriate row, we use the degrees of freedom for the error. In this example our df_e was 12. So, with an alpha level of .05, our Tukey value must be larger than 3.77 before we consider it statistically significant. I know this because the value in Appendix D for 3 groups and 12 degrees of freedom is 3.77.

My Tukey value comparing Groups 1 and 2 was only .95. Because this is smaller than the value of 3.77, I conclude that Groups 1 and 2 did not differ significantly in how much they like bubble gum ice cream, on average. The Tukey values for the comparison of Group 1 with Group 3 produced a Tukey value of 4.29, which is larger than 3.77, so I can conclude that Group 1 is different from Group 3. My third Tukey test produced a value of 3.33, indicating that Group 2 was not significantly different from Group 3. By looking at the means presented for each group in Table 9.3, I can see that, on average, 5-year-olds like bubble gum ice cream more than 12-year-olds, but 5-year-olds do not differ significantly from 8-year-olds and 8-year-olds do not differ significantly from 12-year-olds in how much they like bubble gum ice cream.

WRAPPING UP AND LOOKING FORWARD

One-way ANOVA, when combined with post-hoc tests and *a priori* contrasts, is a powerful technique for discovering whether group means differ on some dependent variable. The *F* value from a one-way ANOVA tells us whether, overall, there are significant differences between our group means. But we cannot stop with the *F* value. To get the maximum information from a one-way ANOVA we must conduct the *post-hoc* tests to determine *which* groups differ from each other. ANOVA incorporates several of the concepts that we have discussed in previous chapters. The sums of squares used in ANOVA is based on the squared deviations first introduced in Chapter 2 when discussing variance. The comparisons of group means is similar to the information about independent samples *t* tests presented in Chapter 8. And the eta-squared statistic, which is a measure of association between the independent and dependent variables, is related to the concepts of shared variance and variance explained discussed in Chapter 7 as well as the notion of effect size discussed in Chapter 6.

In this chapter, a brief introduction to the most basic ANOVA model and post-hoc tests was provided. It is important to remember that many models are not this simple. In the real world of social science research, it is often difficult to find groups with equal numbers of cases. When groups have different numbers of cases, the ANOVA model becomes a bit more complicated. I encourage you to read more about one-way ANOVA models, and offer some references to help you learn more. In the next two chapters, we examine two more advanced types of ANOVA techniques: factorial ANOVA and repeated-measures ANOVA.

In this chapter and those that preceded it, we have examined several of the most basic, and most commonly used, statistics in the social sciences. These statistics form the building blocks for most of the more advanced techniques used by researchers. For example, *t* tests and one-way ANOVA represent the basic techniques for examining the relations between nominal or categorical independent variables and continuous dependent variables. More advanced methods of examining such relations, such as factorial ANOVA and repeated-measures ANOVA are merely elaborations of the more basic methods we have already discussed. Similarly, techniques for examining the relations among two or more continuous variables are all based on the statistical technique already discussed in Chapter 7, correlations. More advanced techniques, such as factor analysis and regression, are based on correlations.

In the remaining chapters of this book, a number of more advanced statistical techniques are described. Because the purpose of this book is to provide a short, non-technical description of a number of statistical methods commonly used by social scientists, there is not adequate space to provide detailed descriptions of these more advanced techniques. Specifically, the technical descriptions of the formulas used to generate these statistics is beyond the scope and purpose of this book. Therefore, in the chapters that follow, general descriptions of each technique are provided including what the technique does, when to use it, and an example of results generated from a statistical analysis using each technique. Suggestions for further reading on each technique are also provided.

GLOSSARY OF TERMS AND SYMBOLS FOR CHAPTER 9

Between group: Refers to effects (e.g., variance, differences) that occur between the members of different groups in an ANOVA.

***F* value:** The statistic used to indicate the average amount of difference between group means relative to the average amount of variance within each group.

Grand mean: The statistical average for all of the cases in all of the groups on the dependent variable.

Mean square between: The average squared deviation between the group means and the grand mean.

Mean square within: The average squared deviation between each group mean and the individual scores within each group.

One-way ANOVA: Analysis of variance conducted to test whether two or more group means differ significantly on a single dependent variable.

Post-hoc tests: Statistical tests conducted after the obtaining the overall *F* value from the ANOVA to examine whether each group mean differs significantly from each other group mean.

Studentized range statistic: Distributions used to determine the statistical significance of post-hoc tests.

Sum of squared between: The sum of the squared deviations between the group means and the grand mean.

Sum of squares error: The sum of the squared deviations between individual scores and group means on the dependent variable.

Sum of squares total: Sum of the squared deviations between individual scores and the grand mean on the dependent variable. This is also the sum of the SS_b and the SS_e.

Tukey HSD: Name of a common post-hoc test.

MS_w	Symbol for the mean square within groups.
MS_e	Symbol for the mean square error (which is the same as the mean square within groups).
MS_b	Symbol for the mean square between groups.
SS_e	Symbol for the sum of squares error (or within groups).
SS_b	Symbol for the sum of squares between groups.
SS_T	Symbol for the sum of squares total.
X_T	Symbol for the grand mean.
F	Symbol for the *F* value.
df	Symbol for degrees of freedom.
K	Symbol for the number of groups.
N	Symbol for the number of cases in all of the groups combined.
n	Symbol for the number of cases in a given group (for calculating SS_b).
N_T	Symbol for the number of cases in each group (for Tukey HSD test).

RECOMMENDED READING

Marascuilo, L. A., & Serlin, R. C. (1988). *Statistical methods for the social and behavioral sciences* (pp. 472-516). New York: Freeman.

Iverson, G. R., & Norpoth, H. (1987). *Analysis of variance* (2nd ed.). Newbury Park, CA: Sage.

CHAPTER 10

FACTORIAL ANALYSIS OF VARIANCE

In the previous chapter, we examined one-way ANOVA. In this chapter and the two that follow, we explore the wonders of two more advanced methods of analyzing variance: factorial ANOVA and repeated-measures ANOVA. These techniques are based on the same general principles as one-way ANOVA. Namely, they all involve the partitioning of the variance of a dependent variable into its component parts (e.g., the part attributable to between-group differences, the part attributable to within-group variance, or error). In addition, these techniques allow us to examine more complex, and often more interesting questions than is allowed by simple one-way ANOVA. As mentioned at the end of the last chapter, these more advanced statistical techniques involve much more complex formulas than those we have seen previously. Therefore, in this chapter and those that follow only a basic introduction to the techniques is offered. You should keep in mind that there is much more to these statistics than described in these pages, and should consider reading more about them in the suggested readings at the end of each chapter.

When to Use Factorial ANOVA

Factorial ANOVA is the technique to use when you have one continuous (i.e., interval or ratio scaled) dependent variable and two or more categorical (i.e., nominally scaled) independent variables. For example, suppose I want to know whether boys and girls differ in the amount of television they watch per week, on average. Suppose I also want to know whether children in different regions of the United States (i.e., East, West, North, and South) differ in their average amount of television watched per week. In this example, average amount of television watched per week is my dependent variable, and gender and region of the country are my two independent variables. This is known as a 2 x 4 factorial analysis, because one of my independent variables has two levels (gender) and one has four levels (region). If I were writing about this analysis in an academic paper, I would write "I conducted a 2 (gender) x 4 (region) factorial ANOVA."

Now when I run my factorial ANOVA, I will get three interesting results. First, I will get two **main effects**, one for my comparison of boys and girls and one for my comparison of children from different regions of the country. These results are similar to the results I would get if I simply were to run two one-way ANOVAs, with one important difference, which I describe in the next section. In addition to these main effects, my factorial ANOVA also produces an **interaction effect,** or simply an **interaction.** An interaction is present when the differences between the groups of one independent on the dependent variable vary according to the level of a second independent variable. Interaction effects are also known as **moderator** effects. I discuss interactions in greater detail in the next section as well. For now, suffice it to say that interaction effects are often very interesting and important pieces of information for social scientists.

Some Cautions

Just as with one-way ANOVA, when conducting a factorial ANOVA it is important to determine whether there are roughly an equal number of cases in each group and whether the amount of variance within each group is roughly equal (known as **homogeneity of variance**). As discussed in the previous chapter, the ideal situation in ANOVA is to have roughly equal sample sizes in each group and a roughly equal amount of variation (e.g., the standard deviation) in each group. If the sample sizes are equal but the standard deviations are significantly different, this is not a big problem. But if the sample sizes *and* the standard deviations are quite different in the various groups, there is a problem. This is particularly likely to be a problem in factorial ANOVA because the combination of independent variables often produces a small number of cases in each group.

In our previous example, suppose we have 20 boys and 20 girls in our entire sample. In addition, suppose that we have 20 children from each of the four regions in our sample. To test the main effects, these numbers are acceptable. That is, it is reasonable to compare 20 boys to 20 girls if we want to know whether boys and girls differ in their average amount of television viewing. Similarly, it is reasonable to compare 20 children from each of the four different regions of the country. But suppose that in the West, our sample of 20 children includes only 5 girls and 15 boys, whereas in the North our sample includes 15 girls and only 5 boys. When we divide up our sample by *two* independent variables, it is easy to wind up with **cell sizes** that are too small to conduct meaningful ANOVAs. A "cell" is a subset of cases representing one unique point of intersection between the independent variables. In the aforementioned example, there would be eight cells: girls from the West, boys from the West, girls from the South, boys from the South, and so on. When you consider that factorial ANOVAs can have more than two independent variables, the sample can be subdivided a number of times. Without a large initial sample, it is easy to wind up with cells that contain too few cases. As a general rule, cells that have fewer than 10 cases are too small to include in ANOVAs, and cell sizes of at least 20 are preferred.

FACTORIAL ANOVA IN DEPTH

When dividing up the variance of a dependent variable, such as hours of television watched per week, into its component parts, there are a number of components that we can examine. In this section, we examine three of these components: The main effects, interaction effects, and **simple effects.** In addition, I also present an introduction to the idea of **partial** and **controlled effects**, an issue that is revisited in Chapter 12 on multiple regression.

Main Effects and Controlled or Partial Effects

As mentioned earlier, a factorial ANOVA will produce main effects for each independent variable in the analysis. These main effects will each have their own *F* value, and are very similar to the results that would be produced if you just conducted a one-way ANOVA for each independent variable on the dependent variable. However, there is one glorious benefit of looking at the main effects in a factorial ANOVA rather than separate one-way ANOVAs: When looking at the main effects from a factorial ANOVA, it is possible to test whether there are significant differences between the groups of one independent variable on the dependent variable while **controlling for**, or **partialing out** the effects of the other independent variable(s) on the dependent variable. Let me clarify this confusing sentence by returning to my example of television viewing.

Suppose that when I examine whether boys and girls differ in the average amount of television they watch per week, I find that there is a significant difference: Boys watch significantly more television than girls. In addition, suppose that children in the North watch, on average, more television than children in the South. Now, suppose that, in my sample of children from the Northern region of the country, there are twice as many boys as girls, whereas in my sample from the South there are twice as many girls as boys. Now I've got a potential problem. How do I know whether my finding that children in the North watch more television than children in the South is not just some *artifact* caused by the greater proportion of boys in my Northern sample? By "artifact" I mean that the North-South difference is merely a *by-product* of the difference between boys and girls; region of the country is not an important factor in and of itself. Think about it: If I already know that boys watch more television, on average, than girls, then I would *expect* my Northern sample to watch more television than my Southern sample because there is a greater proportion of boys in my Northern sample than in the Southern sample. So my question is this: How can I determine whether there is a difference in the average amount of television watched by children in the North and South *beyond* the difference caused by the unequal proportions of boys and girls in the samples from these two regions. Phrased another way, is there an effect of region on television viewing *beyond,* or *in addition to* the effect of gender?

To answer this intriguing question, I must examine the main effect of region on television viewing after *controlling for,* or *partialing out* the effect of gender. I can do this in a factorial ANOVA. To understand how this is accomplished, keep in mind that what we are trying to do with an ANOVA is to *explain* the variance in our dependent variable (amount of television children watch per week) by dividing that variance up into its component parts. If boys and girls differ in how much they watch television, then part of the variance is explained, or accounted for, by gender. In other words, we can understand a bit of the differences among children in their weekly television viewing if we know their gender. Now, once we remove that portion of the total variance that is explained by gender, we can test whether any *additional* part of the variance can be explained by knowing what region of the country children are from. If children from the North and South *still* differ in the amount of television they watch, after *partialing out* or *controlling for* the chunk of variance explained by gender, then we know that there is a main effect of region *independent of* the effect of gender. In statistical jargon, we would say "There is a main effect of region on amount of television watched *after controlling for the effect of gender.*" This is powerful information. In factorial ANOVA, it is possible to examine each main effect and each interaction effect when controlling for *all* other effects in the analysis.

Interactions

A second benefit of factorial ANOVA is that it allows researchers to test whether there are any statistical interactions present. Interactions can be a complex concept to grasp. Making the whole issue even more confusing is that the level of possible interactions increases as the number of variables increases. For example, when there are two independent variables in the analysis, there are two possible main effects and one possible two-way interaction effect (i.e., the interaction between the two independent variables). If there are three independent variables in the analysis, there are three possible main effects, three possible two-way interaction effects, and one possible three-way interaction effect. The whole analysis can get very complicated very quickly. To keep things simple, let's take a look at two-way interactions first.

In my television-viewing example, suppose that I randomly selected 25 boys and 25 girls from each of the four regions of the country, measured the number of hours each child spent watching television, and calculated the averages for each group. (**Note:** Unlike the example provided earlier, in this example there are equal numbers of boys and girls from each region in the sample.) These averages are presented in Table 10.1.

TABLE 10.1   Mean Amounts of Television Viewed by Gender and Region.

	North	*East*	*West*	*South*	*Overall Averages by Gender*
Girls	20	15	15	10	15
Boys	25	20	20	25	22.5
Overall Averages by Region	22.5	17.5	17.5	17.5	

As we see when examining the means in Table 10.1, boys in each region watch more television, on average, than girls. The overall averages by gender presented in the last column indicate that there appears to be a main effect for gender, with boys watching an average of 22.5 hours of television per week and girls watching an average of only 15 hours per week. When we look at the overall averages presented for each region (bottom row), we can see that children in the North watch more television, on average, than do children in the other three regions. Therefore, we can tell that there appear to be main effects for gender and region on amount of television watched. Notice that I said "*appear* to be main effects." To determine whether these main effects of statistically significant, we would have to determine the probability of obtaining differences of this size between randomly selected groups of this size (see Chapter 6 for a discussion of significance tests and their meaning).

Once we have examined the main effects, we can turn our attention to the possible interaction effects. To do this, we need to examine the means in each of the eight cells presented in Table 10.1 (i.e., Northern boys, Northern girls, Eastern boys, Eastern girls, etc.). When we examine these means, we can see that in the North, East, and West, boys watch an average of 5 hours more television per week than girls. But in the South, boys watch an average of 15 more hours of television than girls. Therefore, it appears that the differences in the amount of television watched by girls and boys are not uniform across the four regions of the country. In other words, the relationship between gender and amount of television watched *depends on*, or *is moderated by*, the region of the country. Because the definition of a two-way interaction is that the relationship between an independent variable and a dependent variable is moderated by a second independent variable, we appear to have a two-way interaction here.

When we find a statistically significant interaction (again, we must examine the p value of the F ratio for the interaction term to determine if it is statistically significant), we must determine the *nature* of the interaction and then *describe* the interaction. One excellent method for getting a handle on the nature of the interaction is to depict it graphically. To do this, all we need to do is to graph the means. Line graphs and bar graphs work best. I have produced a line graph that represents the data presented in Table 10.1. This graph is presented in Figure 10.1.

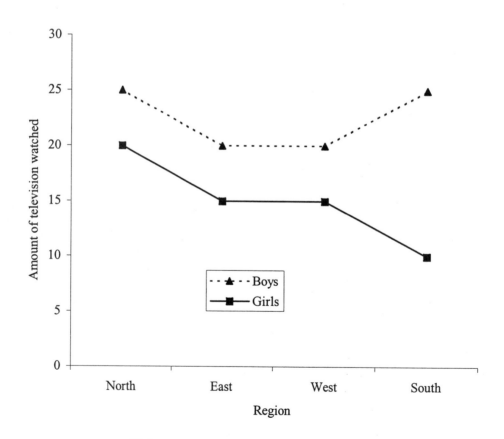

FIGURE 10.1 Interaction of gender and region.

When we look at this graph, the nature of the interaction becomes readily apparent. Specifically, what we can see is that there is a consistent pattern for the relationship between gender and amount of television viewed in three of the regions (North, East, and West), but in the fourth region (South) the pattern changes somewhat. Specifically, the gap between boys and girls in the average amount of television watched per week is much wider in the South than in the other three regions. In Figure 10.1, we can see that the means for boys and girls on the dependent variable are further apart in one region than in the other, but the lines never cross. That is, there is no region in which girls have higher average scores than boys on the dependent variable.

As you look at the graph presented in Figure 10.1, notice that you can see both the main effects and the interaction effects. Recall that the main effect for gender indicates that, when we combine the scores from all four regions, boys appear to have higher average scores than girls on our dependent variable (i.e., amount of television watched per week). In Figure 10.1 this effect is clear, as we can see that the line for boys is higher than the line for girls in all four regions. We can also see evidence of a main effect for region, although this effect is somewhat less clear than the main effect for boys. We see the region effect by noting that for both boys and girls, the average amount of television viewing is higher in the North than in either the East or West regions. This main effect is complicated a bit, however, by the presence of the interaction. Notice that whereas the mean is lower in the South than in the North for girls (supporting our main effect for region), the mean for boys in the South is equal to the mean for boys in the North. This raises a difficult question: When we say there is a main effect for region, with children in the North watching more television, on average, than children in the other three regions, are we being accurate?

Interpreting Main Effects in the Presence of an Interaction Effect

Researchers do not always agree on the best way to interpret main effects when there is a significant interaction effect. Some argue that it makes little sense to interpret main effects at all when there is an interaction effect present, because the interaction effect essentially modifies the meaning of the main effect. In the preceding example, the main effect for region that shows children in the North watch more television than children elsewhere is really only true within the girls sample. In fact, boys in the South watch as much television as boys in the North, and girls in the North do not watch more television than boys in any region. Therefore, some would argue that we should just describe the nature of the interaction, and not interpret the main effects. The logic of this argument is as follows: If I say that children in the North watch more television than children in other regions, the statement is misleading because it is not true for boys. To be accurate, I should just say that *girls* in the North watch more television that *girls* in other regions.

Others, myself included, think it makes sense to interpret all of the effects and to consider them in relation to each other. Returning to our earlier example, we can see that there is a main effect for gender, with boys watching more television, on average, than girls. We can also see that this effect is especially pronounced in the South. In addition, we can say that overall, when we *combine* the samples of boys and girls together, there is a main effect for region such that Northern children watch more television than children in other regions, on average. When we add the consideration of the interaction effect, we can further argue that this overall effect is due *primarily* to differences within the sample of girls, and less to variation within the sample of boys. It is possible to get an interaction effect without a main effect (see Figure 10.2. In this example, boys and girls have equal means, as do children in each of the four geographic regions). Therefore, it makes sense to report and interpret significant main effects, even in the presence of an interaction effect. The key is to provide enough information so that readers of your results can makes sense of them. To do this, it may be necessary to discuss your interaction and main effects in relation to each other.

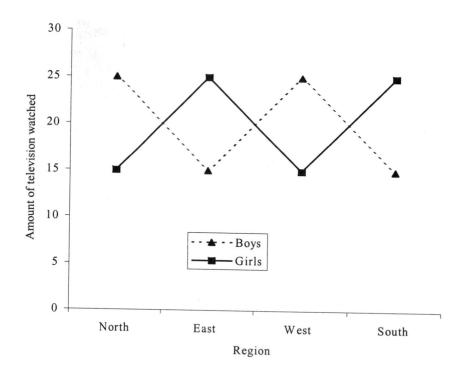

FIGURE 10.2 Interaction with equal means.

Here is another example to more clearly illustrate the problems of interpreting main effects in the presence of significant interactions. Suppose that I were to examine the math skills of boys and girls in two different types of mathematics programs. Students in the "Traditional" program study math in the usual way, reading a textbook and working out math problems in class. Students in the "Experimental" program work in groups to solve problems collaboratively and work with more real-world, applied problems. After one year, I give a math test to 25 randomly selected boys and 25 randomly selected girls from each math program. I calculate the averages for these four groups, which are presented in Figure 10.3.

The means presented in the figure clearly show that although boys and girls in the Traditional math program had similar average scores on the math test, girls did much better than boys in the Experimental math program. This is an interaction. In addition, because girls in the Experimental program did so well on their math test, the overall mean for the Experimental group was significantly higher than the overall mean for the Traditional group, thereby creating a main effect for math program. But does it make sense to say that students in the Experimental math program did better on the test than students in the Traditional program? Clearly, this was not the case for boys, and some would argue that it would be misleading to point out the main effect for math program because the effect is only present for the girls, not the boys.

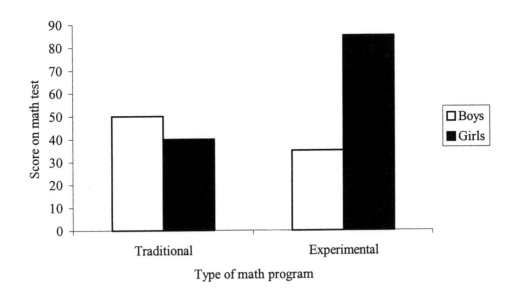

FIGURE 10.3   Interaction of gender by math program.

Testing Simple Effects

Once we have found our main and interaction effects in factorial ANOVA, we can conduct one final set of analyses to examine the simple effects. The methods used to calculate the simple effects and determine whether they are statistically significant are analogous to the post-hoc tests described in Chapter 9. What simple effects analysis allows us to do is to test whether there are significant differences in the average scores of any particular cells. One of the benefits of simple effects analysis is that is allows us to better understand some of the complexities in our data, particularly how to make sense of significant interaction effects.

Returning to our sample data presented in Figure 10.3 we can see that we have four cells: girls in the Traditional math program, Traditional boys, Experimental girls, and Experimental boys. With a simple effects analysis, we can test whether boys in the Traditional math program ($X = 50$) had significantly higher average math test scores than did boys in the Experimental program ($X = 35$). We could also test whether boys and girls in the Traditional program differed significantly. Perhaps most important for helping us understand the interaction effect, we can test whether girls in the Experimental program had higher average math test scores than students in each of the three other groups. For a detailed description of the methods for calculating simple effects, I recommend reading Hinkle, Wiersma, and Jurs (1998).

Analysis of Covariance

Earlier in this chapter, I suggested that one of the benefits of conducting factorial ANOVAs is that it allows us to determine whether groups differ on some dependent variable while controlling for, or partialing out, the effects of other independent variables. A closely related concept that applies to all types of ANOVA, including one-way, factorial, and repeated-measures, is the use of **covariates** in these analyses. In **analysis of covariance (ANCOVA)**, the idea is to test whether there are differences between groups on a dependent variable after controlling for the effects of a different variable, or set of variables. The difference between an ANCOVA and the types of controlled variance I described earlier is that with an ANCOVA, the variable(s) that we are controlling, or partialing out, the effects of is not necessarily an independent variable. Let me explain.

In my earlier example, I was able to test whether boys and girls differed in the amount of television they watched while controlling for the effects of which region of the country they lived in (the second independent variable), as well as the interaction between the two independent variables.

But in an ANCOVA analysis, we can control the effects of variables *besides* independent variables. For example, I could use socioeconomic status (SES) as a covariate and test whether children in different regions of the country differ in the amount of television they watch *after* partialing out the effects of their SES. Suppose that my sample of children from the North is less wealthy than my samples from the three other regions. Suppose further than children from poorer families tend to watch more television than children from wealthier families. Because of this, my earlier results that found greater television watching among children in the Northern region may simply be due to the fact that these children are less wealthy than children in the other regions. With ANCOVA, I can test whether the difference in the viewing habits of children from different regions is due strictly to differences in SES, or whether there are regional differences *independent* of the effects of SES. This is particularly handy because even though factorial ANOVA only allows us to use categorical (i.e., nominally scaled) independent variables, with ANCOVA we can also control the effects of continuous (i.e., intervally scaled) variables.

Effect Size

As I did in Chapter 9, I illustrate effect size in factorial ANOVA, along with some of the particulars about sums of squares, mean squares, and F values, using output from an analysis of my own data using the SPSS computer software program. In this example, students' confidence in their ability to understand and successfully complete their English classwork, referred to here as self-efficacy, was the dependent variable. I wanted to see whether high school boys and girls differed in their self-efficacy (i.e., a main effect for gender), whether students with relatively high grade point averages (GPA) differed from those with relatively low GPAs in their self-efficacy (i.e., a main effect for GPA), and whether there was an interaction between gender and GPA on self-efficacy. Gender, of course, is a two-category independent variable. To make GPA a two-category variable, I divided students into high- and low-GPA groups by splitting the sample in two using the median GPA. This allowed me to perform a 2 (gender) x 2 (GPA) factorial ANOVA. Self-efficacy was measured using a survey with a 5-point scale (1 = "not at all confident" and 5 = "very confident"). My sample consisted of 468 high school students.

The results presented in Table 10.2 begin with descriptive statistics. These statistics are presented separately by subgroups (e.g., low-achieving girls, high-achieving girls, all girls combined, low-achieving boys, etc.). The means and standard deviations presented are for the dependent variable, self-efficacy. By glancing over the means we can see that the boys in our sample had slightly higher average feelings of self-efficacy than did the girls, and this difference appears to be largest among the boys and girls in the low-GPA group.

Turning our attention to the ANOVA results, there are a number of important features to notice. In the far left column titled "Source," there are the various *sources* of variation in self-efficacy. These are the different ways that the variance of the dependent variable, self-efficacy, is sliced up by the independent variables. The first source is called the "Corrected Model." This is the combination of all of the main and interaction effects. If covariates were used, these effects would be included in the "Corrected Model" statistics. Reading from left to right, we can see that the full model has a sum of squares (11.402), which when divided by three degrees of freedom ("*df*") produces a "Mean Square" of 3.801. When we divide this by the mean square error a few rows down (MS_e = .649) we get an F value of 5.854. This has a "Sig." of .001 (in other words, p = .001). Because this value is less than .05 (see Chapter 6), the overall model is statistically significant. But is it *practically* significant? In the final column labeled "Eta Squared" we can see that the overall model accounts for only 3.6% of the variance in self-handicapping scores. In other words, gender, GPA level, and the interaction of these two *combined* only explain 3.6% of the variance. Although this is *statistically* significant, this may not be a big enough effect size to be considered *practically* significant. Remember that statistical significance is influenced by sample size, and 468 cases is a pretty large sample. An effect size of .036, in contrast, is not affected by sample size and therefore may be a better indicator of practical significance.

TABLE 10.2 SPSS results for gender by GPA factorial ANOVA.

Gender	GPA	Mean	Std. Deviation	N
Girl	1.00	3.6667	.7758	121
	2.00	4.0050	.7599	133
	Total	3.8438	.7845	254
Boy	1.00	3.9309	.8494	111
	2.00	4.0809	.8485	103
	Total	4.0031	.8503	214
Total	1.00	3.7931	.8208	232
	2.00	4.0381	.7989	236
	Total	3.9167	.8182	468

ANOVA Results

Source	Type III Sum of Squares	df	Mean Square	F	Sig.	Eta Squared
Corrected Model	11.402	3	3.801	5.854	.001	.036
Intercept	7129.435	1	7129.435	10981.566	.000	.959
Gender	3.354	1	3.354	5.166	.023	.011
GPA	6.912	1	6.912	10.646	.001	.022
Gender * GPA	1.028	1	1.028	1.584	.209	.003
Error	301.237	464	.649			
Total	7491.889	468				
Corrected Total	312.639	467				

In addition to the F value, p value, and effect size for the entire model, SPSS prints out statistics for each of the main effects and the interaction effect as well. Here we can see that gender is a statistically significant predictor of self-efficacy ("Sig." = .023), but the effect size of .011 is very small. By looking at the overall means for girls and boys in the top portion of Table 10.2, we can see that boys ($\overline{X}_{boys}$ = 4.0031) have the slightly higher average feelings of self-efficacy in English than girls ($\overline{X}_{girls}$ = 3.8438). GPA has a larger F value (F = 10.646) and is statistically significant, but it also has a small eta squared value (.022). Students in the high-GPA group had slightly higher average feelings of self-efficacy ($\overline{X}_{high}$ = 4.0381) than did students in the low-GPA group ($\overline{X}_{low}$ = 3.793). The gender by GPA interaction was not statistically significant and has a tiny effect size. Overall, then, the statistics presented in Table 10.2 reveal that although there are statistically significant main effects for gender and GPA on self-efficacy, and the overall model is statistically significant, these effect sizes are quite small and suggest that there is not a strong association between either gender nor GPA with self-efficacy among this sample of high school students.

There are two other features of the SPSS output presented in Table 10.2 worth noting. First, the sum of squares that SPSS uses by default in a factorial ANOVA is called "Type III" sum of squares. This means that when SPSS calculates the sum of squares for a particular effect, it does so by accounting for the other effects in the model. So when the sum of squares for the gender effect is calculated, for example, the effect of GPA and the gender by GPA interaction effects have been partialed out already. This allows us to determine the *unique* effect of each main effect and interaction effect. Second, notice that the *F* value for each effect is obtained by dividing the mean square for that effect by the mean square error. This is the same way *F* values were calculated in one-way ANOVA discussed in Chapter 9.

EXAMPLE: PERFORMANCE, CHOICE, AND PUBLIC VERSUS PRIVATE EVALUATION

In a study published in 1987, Jerry Burger, a psychology professor at Santa Clara University, examined the effects of choice and public versus private evaluation on college students' performance on an anagram-solving task. This experiment involved one dependent variable and two independent, categorical variables. The dependent variable was the number of anagrams solved by participants in a 2-minute period. One of the independent variables was whether participants were able to choose the type of test they would perform. There were 55 participants in the study. About half of these were randomly assigned into the "choice" group. This group was told that they could choose one test to perform from a group of three different tests. The "no choice" group was told that they would be randomly assigned one of the tests. In fact, the "choice" and "no choice" groups worked on the same tests, but the choice group was given the perception that they had chosen the type of test they would work on. So this first independent variable has two categories: Choice and no choice. The second independent variable also had two categories: public versus private. Participants were told either that their test score and ranking would be read aloud, along with their name (the public condition), or that the test scores and ranks would be read aloud without identifying the name of the test taker (the private condition). Participants were randomly assigned to the public-private groups as well. The resulting ANOVA model for this experiment is a 2 (choice vs. no choice) x 2 (public vs. private feedback) factorial ANOVA.

TABLE 10.3   Mean number of anagrams solved for four treatment groups.

	Public		Private	
	Choice	No Choice	Choice	No Choice
Number of anagrams solved	19.50	14.86	14.92	15.36

The average number of anagrams solved by the members of each group are presented in Table 10.3. These means are also graphed in Figure 10.4. Burger found a main effect for the choice independent variable, such that participants who thought they were given a choice of which type of test to take solved more anagrams, on average, than those who were not given a choice. In addition, Burger found that participants in the public evaluation condition solved more anagrams, on average, than participants in the private feedback condition. This is a second significant main effect. Finally, he found an interaction between the two independent variables. If you look closely at the means in Table 10.4 and in Figure 10.4, you can see that three of the four groups have very similar means. Only the public/choice group appears to have solved a significantly greater number of anagrams than did students in the other three groups. I could conduct a test of simple effects to determine whether students in the public/choice group scored significantly higher than students in the public/no-choice group.

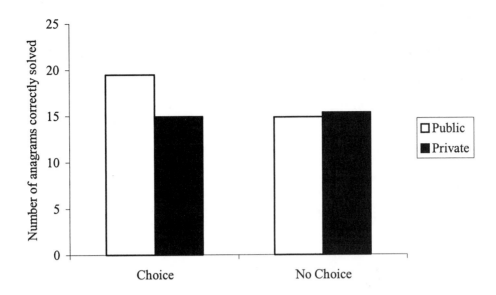

FIGURE 10.4 Interaction of choice by public vs. private evaluation.

In this example, the presence of a significant interaction raises questions about how to interpret our statistically significant main effects. Notice that Burger found a main effect for choice, with students in the two choice groups *combined* solving more anagrams, on average, than students in the two no-choice groups *combined*. The problem here is that we can see that students in the private/choice group did not score higher than students in the private/no-choice group, and had very similar scores to students in the public/no-choice group. Therefore, this main effect for choice versus no choice is caused entirely by the relatively high scores of the public/choice group. So when Burger states that participants solved more anagrams on average when they were given a choice than did participants who had no choice, he must carefully point out that this is only true for students in the public condition. Similarly, the main effect for public over private is also caused solely by the high scores of the public/choice group. By noting that there is a significant interaction of the two independent variables, Burger is in effect telling his readers that they must interpret the main effects very carefully. If we were simply to conclude that students perform better when given a choice, or when their performance is made public, we would miss the intricacy of the story.

WRAPPING UP AND LOOKING FORWARD

In this chapter we were able to extend what we learned about ANOVA in Chapter 9 in three important ways. First, we added the concept of *multiple independent variables*. By having more than one independent variable in the model, we were able to more finely divide up, and explain, the variance in the dependent variable. Second, we examined the concept of *controlling* or *partialing out* the effects of other variables in the model, including covariates, to get a better picture of the *unique* relation between an independent and a dependent variable. Finally, in this chapter we were able to consider the importance of statistical interactions. All three of these concepts provide a hint of the amazing power of many different statistical techniques to explore the relations among variables. In the social sciences, as in most fields, variables are related to each other in very complex ways. We live in a complex world. Although *t* tests and one-way ANOVA are useful statistical techniques, they are often unable to examine the most interesting questions in the social sciences. It is the messy world of interactions, shared variance, and multiple predictors that make the statistical life a life worth living. So although the concepts in these last few chapters may seem a bit more difficult than those discussed earlier in the book, they pay rich dividends when finally understood. In the next chapter we enter the complex yet particularly interesting world of repeated-measures ANOVA.

GLOSSARY OF TERMS AND SYMBOLS FOR CHAPTER 10

Analysis of covariance (ANCOVA): An analysis of variance conducted with a covariate. It is an analysis conducted to test for differences between group means after partialing out the variance attributable to a covariate.

Cell size: The number of cases in each subgroup of the analysis.

Covariate(s): A variable, or group of variables, used to control, or account for, a portion of the variance in the dependent variable, thus allowing the researcher to test for group differences while controlling for the effects of the covariate.

Factorial ANOVA: An analysis of variance with at least two categorical independent variables.

Homogeneity of variance: An assumption of all ANOVA models that there are not statistically significant differences in the within-group variances on the dependent variable between the groups being compared.

Interaction (effect): When the relationship between the dependent variable and one independent variable is moderated by a second independent variable. In other words, when the effect of one independent variable on the dependent variable differs at various levels of a second independent variable.

Main effects: These are the effects for each independent variable on the dependent variable. In other words, differences between the group means for *each* independent variable on the dependent variable.

Moderator: When the relationship between the dependent variable and one independent variable differs according to the level of a second independent variable, the second independent variable acts as a moderator variable. It is a variable that moderates, or influences, the relationship between a dependent variable and an independent variable.

Partial, controlled effects: When the shared, or explained variance between a dependent variable and an independent variable (or a covariate) is held constant, thereby allowing the researcher to examine group differences *net of* the controlled effects.

Simple effects: The differences between the means of each subgroup in a factorial ANOVA are called the simple effects. A subgroup involves the division of an independent variable into smaller groups. For example, if ethnicity is one independent variable (African-American, Asian-American, and Hispanic-Latino), and gender is another variable, then each ethnic group has two subgroups (e.g., African-American females and African American-males).

REFERENCES AND RECOMMENDED READING

Burger, J. M. (1987). Increased performance with increased personal control: A self-presentation interpretation. *Journal of Experimental Social Psychology, 23,* 350-360.

Iverson, G. R., & Norpoth, H. (1987). *Analysis of variance* (2nd ed.) Newbury Park, CA: Sage.

Wildt, A. R., & Ahtola, O. T. (1978). *Analysis of covariance.* Newbury Park, CA: Sage.

Hinkle, D. E., Wiersma, W., & Jurs, S. G. (1998). *Applied statistics for the behavioral sciences* (4th ed.). Boston: Houghton Mifflin.

CHAPTER 11

REPEATED-MEASURES ANALYSIS OF VARIANCE

One type of *t* test discussed in Chapter 8 was the **paired *t* test**. One type of study in which a paired *t* test would be used is when we have two scores for a single group on a single measure. For example, if we had a group of third-grade students and we gave them a test of their math abilities at the beginning of the school year and again at the end of the school year, we would have one group (third graders) with two scores on one measure (the math test). In this situation, we could also use a **repeated-measures analysis of variance (ANOVA)** to test whether students scores on the math test were different at the beginning and end of the academic year.

Repeated-measures ANOVA has a number of advantages over paired *t* tests, however. First, with repeated-measures ANOVA, we can examine differences on a dependent variable that has been measured at more than two time points, whereas with an independent *t* test we can only compare scores on a dependent variable from two time points. Second, as discussed in Chapter 10 on factorial ANOVA, with a repeated-measures ANOVA we can control the for effects of one or more covariates, thereby conducting a repeated-measures analysis of covariance (ANCOVA). Third, in a repeated-measures ANOVA, we can also include one or more independent categorical, or **group variables**. This type of mixed model is a particularly useful technique and is discussed in some detail later in the chapter.

When to Use Each Type of Repeated-Measures Technique

The most basic form of a repeated-measures ANOVA occurs when there is a single group (e.g., third graders) with two scores (e.g., beginning of the year, end of the year) on a single dependent variable (e.g., a mathematics test). This is a very common model that is often used in simple laboratory experiments. For example, suppose I wanted to know whether drinking alcohol affects the reaction time of adults when driving. I could take a group of 50 adults and test their reaction time when driving by flashing a red light at each one of them when they are driving and measuring how long it takes for each one to apply the brakes. After calculating the average amount of time it takes this group to apply the brakes when sober, I could then ask each member of my group to consume two alcoholic drinks and then again test their reaction time when driving, using the same methods. In this example, I've got one group (50 adults) with two scores on one dependent variable (reaction time when driving). After the second measure of reaction time, I could ask each of my participants to consume two more alcoholic drinks and again test their reaction time. Now I've got three measures of reaction time that I can use in my repeated-measures ANOVA. Notice that my dependent variable is always the same measure (reaction time), and my group is always the same (sample of 50 adults). The results of my repeated-measures ANOVA will tell me whether, on average, there are differences in reaction time across my three trials. If there are, I might logically conclude that drinking alcohol affects reaction time, although there may be other explanations for my results (e.g., my participants may be getting tired or bored with the experiment, they may be getting used to the test situation, etc.).

In a slightly more advanced form of the test of reaction time, I could include a covariate. In the previous example, suppose that I found the reaction time was fastest when my participants were sober, a bit slower after two drinks, and a lot slower after four drinks. Suppose that I publish these results and the national beer, wine, and liquor companies become worried that, because of my study, people will stop drinking their products for fear of getting in automobile accidents. These producers of alcoholic drinks begin to criticize my study. They suggest that because equal amounts of alcohol generally have greater effects on those who weigh less than on heavier people, my results may have been skewed by the effects of alcohol on the lighter people in my study. "Although the effects of two alcoholic drinks may impair the reaction time of lighter people, even four alcoholic drinks will not impair the reaction time of heavier people" said the United Alcohol Makers of America (a fictitious group).

Stung by the criticism of the UAMA, I decide to replicate my study, but this time I use weight as a covariate. Again, I measure participants' reaction time when driving completely sober, after two alcoholic drinks, and after four alcoholic drinks. In addition, this time I weigh each of my participants. Now when I analyze my data, I include my weight covariate. I find that, after controlling, or partialing out the effects of weight, there is no difference in the reaction time of participants before they have any drinks and after they have two drinks, but after four drinks my participants react more slowly, on average, than they did after zero or two drinks. These results suggest that drinking may increase the reaction time of lighter people after only two drinks, but it seems to impair the reaction time of people, regardless of weight, after four drinks.

Still bothered by my results, the UAMA suggests that my results are skewed because I did not look at the effects of drinking on reaction time separately for men and women. "Women are more dramatically affected by alcohol than men, regardless of weight" claims the UAMA. They argue that although consuming four alcoholic drinks may slow the reaction time of women, it will not have an effect on heavy men. Though I am dubious of the argument that heavy men should have their rights to drink and drive protected, in the name of science I decide to conduct one final study. In this study, again with 50 adults (25 women and 25 men) of various weights, I again test their reaction time while driving after zero, two, and four alcoholic drinks. Now I've got one dependent variable (reaction time) measured at three time points, one covariate (weight), and one independent group variable (gender of participant). Notice that although number of drinks is technically an independent variable, it is not a categorical, or group, variable. In other words, I do not have three independent groups (the zero-drink group, the two-drink group, and the four-drink group). Rather, I have three *dependent,* or repeated, measures of the same dependent variable, reaction time.

When I examine the results of my study, I find that, after controlling for the effects of my covariate (weight), there is still no difference in reaction time measured after zero and two drinks, but still slower reaction time, on average, after four drinks. In addition, I find no interaction between gender and number of drinks on reaction time. This tells me that *both* men and women have slower reaction time after four drinks, regardless of their weight.

To summarize, my three different repeated-measures ANOVAs produced the following results. The first one found that adults' reaction times while driving were slower, on average, after two drinks and slower still after four drinks. My second test included the covariate of weight, and I found that when we control for the effects of weight, reaction time is not slower after two drinks but is slower after four drinks. Finally, in my third analysis, I examined whether changes in reaction time after two and four drinks, when controlling for weight, was different for men and women. I found that it was not. These three analysis provide a snapshot of how repeated-measures ANOVA works and what information it can provide. Now let's take a closer look at how it works.

REPEATED-MEASURES ANOVA IN DEPTH

Repeated-measures ANOVA is governed by the same general principles as all ANOVA techniques. As with one-way ANOVA and factorial ANOVA, in repeated-measures ANOVA we are concerned with dividing up the variance in the dependent variable. Recall that in a one-way ANOVA, we separated the total variance in the dependent variable into two parts: that attributable to differences between the groups, and that attributable to differences among individuals in the same group (a.k.a., the error variance). In a repeated-measures ANOVA with no independent group variable, we are still interested in the error variance. However, we also want to find out how much of the total variance can be attributed to **time**, or **trial**. That is, how much of the total variance in the dependent variable is attributable to differences *within individuals* across the times they were measured on the dependent variable.

Consider an example. Suppose that I am interested in examining whether a group of students increase their knowledge and skills from one academic year to the next. To do this, I give my sample a standardized test of vocabulary (with a possible range of 1-100), once when they are finishing third grade and again when they are finishing fourth grade. When I do this, suppose I get the data presented in Table 11.1.

TABLE 11.1 Vocabulary test scores at two time points.

Case Number	Test Score, Time 1 (Third Grade)	Test Score, Time 2 (Fourth Grade)
1	40	60
2	55	55
3	60	70
4	40	45
5	75	70
6	80	85
7	65	75
8	40	60
9	20	35
10	45	60
Trial (or Time) Average	$X = 51.5$	$X = 61.6$

For each of the 10 cases in Table 11.1, we have two test scores, giving us a total of 20 scores in the table. We could find an average for these 20 scores, and a standard deviation, and a variance. In a repeated-measures ANOVA, we want to try to partition this total variance into different pieces. In the most basic form of repeated-measures ANOVA, there are three ways that we can slice up this variance. First, there is the portion of variance attributable to deviations in scores between the individual cases in the sample. For each case in our sample, we have two scores (one for Time 1 and one for Time 2). We can find an average of these two scores, for each individual, and then see how much this individual average differs from the overall average. In Table 11.1, for example, the first case has an average score of 50 across the two trials (40 + 60   2 = 50). The overall average for the scores in the table if 56.75. So there is some variation in the average scores of the 10 individuals in the sample. This is one source of variation.

The second source of variation in the scores involves the **within-subject variance**, or differences, between Time 1 and Time 2 scores. As we can see by looking at the scores in Table 11.1 and in Figure 11.1, it appears that students generally had different scores on the test at Time 1 than they did at Time 2. These intra-individual, or within-subject, differences between Time 1 and Time 2 scores can be seen more easily in the graph presented in Figure 11.1. These intra-individual changes reflect differences, or variance, *within each individual,* and therefore are called within-subject effects. What we are interested in is whether, on average, individuals' scores were different at Time 1 (in third grade) than they were at Time 2 (in fourth grade). Notice that we are asking whether there were differences in the scores between Time 1 and Time 2 *on average.* If the scores of some of the cases went up from Time 1 to Time 2, but the scores of other cases went down by the same amount, then these changes would cancel each other out, and there would be no *average* difference between the Time 1 and Time 2 scores. But if the scores either went up or down on average between Time 1 and Time 2, then we could say that some of the total variation can be attributable to within-subject differences across time. A look at the scores in Table 11.1 and Figure 11.1 reveals that scores *appeared* to increase from Time 1 to Time 2. To examine whether there are differences in the average scores across time, all we need to do is calculate the average score at each time and find the difference between these average scores and the overall average. In the preceding paragraph, we found that the overall average score was 56.75. In Table 11.1, we can see that the average for Time 1 is 51.5, and the average score for Time 2 is 61.5. So we can see that there is some variance in the average scores at the two times (i.e., third and fourth grade), suggesting that there may be a within-subjects effect.

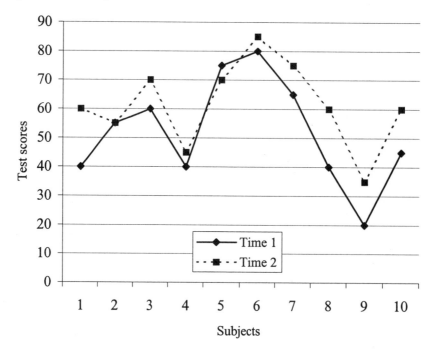

FIGURE 11.1 Time 1 and Time 2 test scores.

The third source of variation in the scores comes from the interaction between the within-subject scores and the variance in scores across the subjects. Although it appears that the scores of the members in our study increased, *on average*, from Time 1 to Time 2, it does not appear that these within-subject changes over time were the same across all of the subjects in the sample. As Figure 11.1 clearly shows, some subjects had large increases in their test scores from Time 1 to Time 2 (e.g., Subjects 1, 8, 9, and 10), whereas others had more modest increases, one had no change (Subject 2) and one actually had a lower score at Time 2 (Subject 5). So there appears to be a case, or subject, by time interaction. In other words, the size of the increase in test score from third to fourth grade depended on the which subject we were looking at. This difference among the subjects in the magnitude of change from Time 1 to Time 2 represents the third source of variance.

Using these three sources of variance, we can then calculate an F ratio and determine whether there are statistically significant differences in the average scores at Time 1 and the average scores at Time 2. To do this, we divide the **mean square for the differences between the trials**, or time, averages (MS_T) by the **mean square for the subject by trial interaction** ($MS_{s \times T}$). The degrees of freedom F ratio is the number of trials minus 1 ($T - 1$) and ($T - 1$)($S - 1$), where S represents the number of subjects in the sample. What we get when we calculate this F ratio is an answer to the following question: How large is the difference between the average scores at Time 1 and Time 2 relative to (i.e., divided by) the average amount of variation among subjects in their change from Time 1 to Time 2? Because differences in the rate of change across time are just considered random fluctuations among individuals, this F ratio, like all F ratios, is a measure of *systematic* variance in scores divided by *random* variance in scores. (Note: For a more detailed discussion of these sources of variance, including how to calculate the sum of squares for each source, see Glass and Hopkins, 1996.)

In this most basic form of repeated-measures ANOVA, notice that what we are primarily concerned with is whether there is a systematic pattern of differences *within individuals, or subjects*, in the scores on the dependent variable measured at two time points. Also notice that if we had three points of data (e.g., test scores from third, fourth, and fifth grades), our question would remain the same: Is there a pattern of differences in the scores *within subjects* over time? Keep in mind that when I say "a pattern" or "a systematic pattern," I mean *on average*. So, to rephrase the question, a simple repeated-measures ANOVA can help us detect whether, *on average,* scores differ *within subjects* across multiple points of data collection on the dependent variable. This type of simple repeated-measures ANOVA is sometimes referred to as a **within-subjects design**.

Repeated-Measures Analysis of Covariance (ANCOVA)

A slightly more complicated form of repeated-measures ANOVA can be produced by adding one or more covariates to the model. As discussed earlier in this chapter, as well as in Chapter 10 on factorial ANOVA, covariates can be used to partial out some portion of the variance in the dependent variable. I illustrate how this works by returning to the example data presented in Table 11.1.

One could argue that the results of my repeated-measures ANOVA were skewed by the scores of the more intelligent students in my sample. Although the students in my sample scored higher on the vocabulary test at the end of fourth grade than they did at the end of third grade, we must keep in mind that the change in scores over time represents an *average* change. Some students in my sample improved quite a bit over time, whereas others did not increase at all, and one (Subject 5) actually declined. So it is possible that this overall average improvement over time was caused by large increases among the brightest students. To explore this hypothesis, I conduct a new repeated-measures ANOVA, but this time I include a covariate: IQ test scores.

When I conduct my repeated-measures ANCOVA, I now have three ways of partitioning the total variance in my vocabulary test scores. First, there is the portion of variance that is accounted for by my covariate, IQ test scores. If students' IQ test scores are related to (i.e., correlated with) their vocabulary test scores, then the IQ test scores will explain, or account for, some percentage of the variance in students' vocabulary test scores (see Chapter 7 for a more thorough explanation of this concept). Second, after partialing out the portion of variance attributable to IQ test scores, I can see whether any of the remaining variance in vocabulary test scores is accounted for by *changes* in vocabulary test scores from third to fourth grade. In other words, once we control for the effects of IQ test scores, do the scores of my sample change significantly from Time 1 (third grade) to Time 2 (fourth grade), on average? Is there still a *within-subjects* effect after controlling for IQ test scores? Finally, after accounting for the variance in vocabulary test scores that is attributable to the covariate (i.e., IQ test scores) and the within-subjects effect (i.e., changes from third to fourth grade), there will still be some variance in vocabulary test scores that is not explained. This is error variance, which is the same as the random variance that we normally find between different members of the same sample.

To reiterate, when a covariate (or several covariates) are added to the repeated-measures ANOVA model, they are simply included to "soak up" a portion of the variance in the dependent variable. Then, we can see whether there are any within-subject differences in the scores on the dependent variable, when *controlling for,* or *partialing out* that portion of the variance accounted for by the covariate(s). In the example we have been using, the addition of the IQ score covariate allows us to answer this question: Do students' vocabulary test scores change, on average, from third to fourth grade *independently* of their IQ scores? Phrased another way, we can ask whether, when controlling for IQ, students' vocabulary test scores change from third to fourth grade.

Adding an Independent Group Variable

Now that we have complicated matters a bit by adding a covariate to the model, let's finish the job by adding an independent categorical, or *group* variable. Suppose, for example, that my 10 cases listed in Table 11.1 included an equal number of boys and girls. This two-level independent variable may allow us to divvy up the variance in our dependent variable even more, but only if there are differences in the scores of boys and girls.

TABLE 11.2   Vocabulary test scores at two time points.

Case Number	Test Score, Time 1 (Third Grade)	Test Score, Time 2 (Fourth Grade)
Boys		
9	20	35
4	30	45
8	40	60
1	40	60
10	45	60
Girls		
2	55	55
3	60	70
7	65	75
5	75	70
6	80	85

There are two ways that this independent group variable may explain variance in vocabulary test scores. First, boys and girls may simply differ in their average scores on the vocabulary test. Suppose that when we divide the scores in Table 11.1 by gender, we get the results presented in Table 11.2. If the data were aligned in this way, we would find that at Time 1, the average score on the vocabulary test was 35 for boys and 65 for girls. Similarly, at Time 2, the average score for boys was 52 whereas for girls it was 71. At both Time 1 and Time 2, boys appear to have lower average scores on the vocabulary test than girls. Therefore, there appears to be a *main effect* for gender. Because this main effect represents a difference between groups of cases in the study, this type of effect is called a **between-groups** or **between-subjects** main effect. In other words, some of the variance in vocabulary test scores can be explained by knowing the group (i.e., gender) to which the student belongs.

The second way that my independent group variable can explain some of the variance in my dependent variable is through an interaction effect. If I were to graph the means for boys and girls at both time points, I would get an interesting picture. As we can see in Figure 11.2, the main effect for gender is clear. In addition, it is also clear that there is a within-subject effect, because both boys and girls have higher scores at Time 2 than they did at Time 1. But what also becomes clear in Figure 11.2 is that the *amount* of change from Time 1 to Time 2 appears to be greater for boys than for girls. Whereas the average score for girls increased 6 points from third to fourth grade, it grew 17 points for boys. These different amounts of change represent another source of explained variance in vocabulary test scores: the interaction of the within-subjects effect with the *between-subjects* effect. In other words, there appears to be a gender (i.e., between-subjects) by time (i.e., within-subjects) interaction on vocabulary test scores.

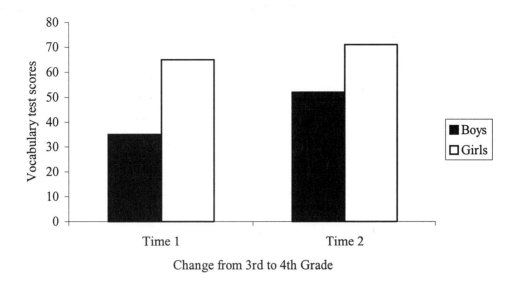

FIGURE 11.1 Gender by time interaction.

To summarize, our final model has a number of effects, each of which can explain some of the variance in the vocabulary test scores of the cases in my sample. First, some of the variance in vocabulary test scores can be explained by students' IQ test scores. On average, students with higher IQ test scores had higher vocabulary test scores. Second, even after controlling for IQ, there was a within-subjects main effect. That is, I can know something about students' scores on the vocabulary test by knowing whether we are talking about Time 1 or Time 2 test scores because, on average, students had higher scores at Time 2. Third, there was a between-subjects effect for gender, so I can explain some of the variance in vocabulary test scores by knowing the gender of the student. Girls had higher scores, on average, than did boys. Fourth, my time by gender interaction explains some additional variance in vocabulary test scores. Although both boys' and girls' scores improved over time, this improvement was more dramatic among boys, on average. Finally, there is some variance in vocabulary test scores that I cannot explain with my covariate, time, gender, or interaction effects. This is error variance.

Please keep in mind that my different effects (time, gender, interaction, covariate) will only explain variation in my dependent variable if the relations between my dependent variable and these effects are statistically significant (see Chapter 6). With only 10 cases in my sample, many of these effects may not be statistically significant.

EXAMPLE: CHANGING ATTITUDES ABOUT STANDARDIZED TESTS

Every year, students across the country take standardized tests of achievement. A few years ago I conducted a study to explore students' beliefs and attitudes about the taking a standardized test, the Iowa Test of Basic Skills (ITBS). The participants in the study included fifth graders from an elementary school and seventh and eighth graders from two middle schools. There were 570 students in the sample. Students were given a survey the week before they took the ITBS and then were given another survey during the week after they took the test. This pretest-posttest design allowed me to examine how students were thinking about the test before taking it, and then to reassess their thinking soon after taking the test.

The two surveys contained questions about a variety of beliefs and attitudes, including test anxiety, self-concept, attributions for success and failure, and other variables related to motivation. One set of questions assessed students' perceptions about the validity of the test. On the pretest survey, the measure of validity beliefs included items such as "I think the ITBS test will be a good measure of what I can do in school" and "The ITBS test will measure how smart I am." On the posttest survey, the measure of validity beliefs included such items as "My score on the ITBS test will tell me how smart I am" and "The ITBS test was a good test of how much I have learned in school." Students answered each of these questions using an 8-point scale ranging from 1 ("strongly disagree") to 8 ("strongly agree"). Students' answers to each question were averaged to create a single pre-test validity score (VALID) and a single post-test validity score (PSTVALID), each with a range of 1 to 8.

One question that we can ask with these data are whether students' beliefs about the validity of the tests, in terms of the test measuring what they know or can do academically, is whether these beliefs changed, on average, from before they took the test to after. Students may develop a set of beliefs about the test before they take it, perhaps due to what their teachers and school administrators tell them in preparation for the test. But once they take the test, and see what sorts of questions the test contains, they may change their beliefs about what the test really measures. This is a *within-subjects* type of question: Are there changes *within individuals* in attitudes about the validity of the test from Time 1 to Time 2?

One factor that may cause students to change their attitudes about the validity of the ITBS test is how well they performed on the test. When taking the test, those who thought the test was difficult, and knew that they were not doing well on it, may develop a somewhat defensive perception that the tests are unfair or invalid. On the other hand, those who felt the test was easy and knew they were doing well when taking the test may tend to develop self-augmenting perceptions of the test, such as the test reveals their intelligence and is a valid measure. To control for these performance-based differences in perceptions of test validity, I add two covariates to the model, both measures of actual test performance. One covariate is the students' scores, in percentile terms, on the math portion of the ITBS test. The other covariate is students' percentile scores on the verbal portion of the test. The addition of these two variables turns my repeated-measures ANOVA into a repeated-measures ANCOVA. This repeated-measures ANCOVA can be used to answer the following question: When controlling for actual achievement on the test, are there changes *within individuals* in students' attitudes about the validity of the test from Time 1 to Time 2?

Finally, it is possible that boys' and girls' perceptions of the validity of the test may differ. Perhaps one gender is more trusting of standardized measures than the other. In addition, perhaps one gender tends to have more idealized perceptions of the tests' validity before taking the test, but these perceptions change after actually taking the test. The other gender, with no such idealized preconceptions, may not change their attitudes after taking the test. By adding the independent group variable of gender, I can now address all of the following questions with my model:

1. When controlling for the effects of gender and achievement, are there changes *within subjects* in students' attitudes about the validity of the test from Time 1 to Time 2?
2. When controlling for within-subject effects and achievement, are their differences between boys' and girls' average beliefs about the validity of the test (i.e., *between-subjects* effects)?
3. Is there a within-subject by between-subject interaction, such that the size of the change in perceptions about the validity of the tests from Time 1 to Time 2 is different for boys and girls, when controlling for the effects of achievement?

As you can see, there are a number of very interesting questions that I can examine in a single repeated-measures ANCOVA. To examine these questions, I conducted my analysis using SPSS software. The actual SPSS output from the analysis is presented in Table 11.3. I explain each piece of information in the order it appears in Table 11.3.

The first set of information in Table 11.3 are the means, standard deviations, and sample sizes for the pretest dependent variable (Pretest Validity) and the post-test dependent variable (Posttest Validity). A quick glance at the separate means for boys and girls on the Pretest Validity and Posttest Validity variables reveals that whereas the girls' averages are virtually identical from Time 1 to Time 2, the boys' mean declines somewhat (from 6.2852 to 6.0076). We can also see that at both Time 1 and Time 2, boys appear to score higher, on average, than girls on the validity perception measures. Whether these differences are statistically significant is still to be determined. Regardless of whether these differences are statistically significant, they may not be *practically* significant: Boys and girls do not appear to differ much in their average perceptions of the validity of the ITBS test.

TABLE 11.3   SPSS output for repeated-measures ANCOVA.

Descriptive Statistics

	Gender	Mean	Std. Deviation	N
Pretest Validity	girl	5.7679	1.5762	307
	boy	6.2852	1.4761	264
	Total	6.0071	1.5510	571
Posttest Validity	girl	5.7096	1.5190	307
	boy	6.0076	1.5324	264
	Total	5.8473	1.5311	571

Tests for Between-Subjects Effects

Source	Type III Sum of Squares	df	Mean Square	F	Sig.	Eta Squared
Intercept	10642.913	1	10642.913	2930.419	.000	.838
Reading Test Score	35.006	1	35.006	9.639	.002	.017
Math Test Score	5.266	1	5.266	1.450	.229	.003
Gender	**41.941**	**1**	**41.941**	**11.548**	**.001**	**.020**
Error	2059.273	567	3.632			

Tests involving within-subjects effects

Source	Type III Sum of Squares	df	Mean Square	F	Sig.	Eta Squared
Validity	**8.884**	**1**	**8.884**	**10.617**	**.001**	**.018**
Validity * Read Test	.164	1	.164	.196	.659	.000
Validity * Math Test	3.533	1	3.533	4.222	.040	.007
Validity * Gender	**3.670**	**1**	**3.670**	**4.386**	**.037**	**.008**
Error(Validity)	474.437	567	.837			

Below the means and standard deviations in the SPSS output, we can find the "Tests for Between-Subjects Effects." Here we can see five separate sums of squares (*SS*), degrees of freedom (*df*), and mean squares (*MS*). We also get to *F* values, "Sig." *p* values, and our effect size measure, "Eta Squared." The statistic we are most interested in here is the *F* value "Sig" *p* value, and "Eta Squared" effect size for the analysis involving Gender. These statistics tell us whether, on average, boys and girls differ in their average perceptions of validity of the ITBS, when controlling for their performance on the test. It is important to remember that this between-subjects test is for the pretest Validity scores and the posttest Validity scores *combined*. Because the "Sig" is a *p* value, and this *p* value is well less than .05, we conclude that, on average, *across times*, boys and girls differ in their perceptions of the validity of the tests. If we take a look at the means presented earlier, we can conclude that boys have more faith in the validity of the test scores than do girls, even after controlling for performance on the test. Notice that the eta squared statistic for the gender effect is quite small (eta^2 = .02), indicating that gender accounts for only 2% of the variance in the combined pretest and posttest Validity scores. This suggests that our *statistically* significant result may not be *practically* significant. The data presented in this part of the table also reveal that there is a significant relationship between one of our covariates (Reading test scores) and our dependent variable (the combined pretest and posttest Validity scores).

Continuing down the SPSS output in Table 11.3, we get to the section labeled "Test involving Within-Subject Effect." Here we are most interested in the results for Validity and the Validity * Gender interaction. Validity is the name that I have given to the combination of the pre- and post-test scores on the validity measure. When these two scores are combined to create a *within-subjects* factor (which I called Validity), we can conduct a test to see whether there were statistically significant within-subject changes, on average, on the validity measures from Time 1 to Time 2. Because this within-subjects test is concerned with *changes* or *differences* within subjects across the two times, the dependent variable in this analysis is *not* the combined scores on the pretest Validity and posttest Validity variables, as it was in the between-subjects test. Rather, the dependent variable is the *difference* or *change* in the scores, *within-subjects,* from Time 1 to Time 2. Because our F value for Validity has a corresponding p value of $p = .001$ (as listed in the "Sig." column), we can see that, on average, students' belief in the validity of the test did change from Time 1 to Time 2. By looking at the means reported earlier, we can see that, on average, students had more faith in the validity of the test *before* they took the test than they did *after* taking the test. The eta-squared statistic for this effect ($eta^2 = .018$) indicates that there was a small effect size for this effect. We can also see, from the Validity * Math Test ($F = 4.222$) and the associated p value ("Sig." = .040), that there was a significant relationship between the math test covariate and our dependent variable in this analysis. In other words, there was significant relationship between how well students performed on the math portion of the ITBS test and how much their beliefs in the validity of the test changed over time. There was no significant relationship between performance on the reading portion of the ITBS test and changes in beliefs about the validity of the test.

In addition to the main within-subject effect, we can see that there is a significant Validity by Gender interaction ("Sig.", or $p = .037$). This tells us that the within-subject changes from Time 1 to Time 2 in beliefs about the validity of the ITBS test were not of equal size among boys and girls. If you recall from the means presented at the top of Table 11.3, this comes as no surprise. We can see that whereas girls' mean score on the validity variable changed little from Time 1 to Time 2, for boys there was a noticeable decrease in beliefs about the validity of the test from Time 1 to Time 2. It is important to keep in mind that even the statistically significant results in this analysis are all quite modest, as revealed by the small effect sizes (see Chapter 6 for a discussion of effect size).

Now that we have found a significant interaction, we perhaps need to modify our conclusions about the main effects we have found. First, the differences between boys' and girls' average perceptions that the test is valid appear to be due primarily to the relatively large gap in Time 1 scores. Boys' and girls' perceptions of the validity of the test were more similar after they actually took the test, although boys were still slightly more likely to believe the tests were valid. Second, the statistically significant within-subject change in beliefs about test validity over time appear to be caused entirely by changes in the boys' perceptions from Time 1 to Time 2. Girls barely changed their beliefs about validity at all over time.

Taken as a group, the results of our repeated-measures ANCOVA reveal a great deal about how boys and girls think about the validity of the ITBS. First, we know that although performance on the English portion of the test is related to beliefs about the validity of the test, it is performance on the math portion of the test that is related to *changes* in beliefs about validity. Second, we know that boys tend to view the tests as more valid than girls, particularly before they take the test, regardless of how well students performed on the test (i.e., controlling for the effects of test scores). Third, we know that students tend to decline in their beliefs about the validity of the test after taking the test, but this decline appears to only occur among boys. Finally, we know that all of these effects are quite small because the small effect sizes tell us so. This is a lot of information, and it demonstrates the power of repeated measures.

WRAPPING UP AND LOOKING FORWARD

In some ways, repeated-measures ANOVA is a simple extension of ideas we have already discussed. The similarities with paired t tests (Chapter 8) are clear, as is the idea of parsing up the variance of a dependent variable into various components. But the tremendous power of repeated-measures ANOVA can only be appreciated when we take a moment to consider all of the pieces of information that we can gain from a single analysis. The combination of within-subjects and between-subjects variance, along with the interaction between these components, allows social scientists to examine a range of very complex, and very interesting, questions. Repeated-measures ANOVA is a particularly useful technique for examining change over time, either in longitudinal studies or in experimental studies using a pre-treatment, post-treatment design. It is also particularly useful for examining whether patterns of change over time vary for different groups.

In the final chapter of the book we will examine one of the most widely used and versatile statistical techniques: regression. As you finish this chapter and move onto the next, it is important to remember that we are only able to scratch the surface of the powerful techniques presented in the last three chapters of this book. To gain a full appreciation of what factorial ANOVA, repeated-measures ANOVA, and regression can do, you will need to read more about these techniques.

GLOSSARY OF TERMS AND SYMBOLS FOR CHAPTER 11

Between-subjects effect: Differences attributable to variance among the scores on the dependent variable for individual cases in the ANOVA model.

Between-groups effect: Differences in the average scores for different groups in the ANOVA model.

Group variable(s): Categorical independent variables in the ANOVA model.

Mean square for the differences between the trials: The average squared deviation between the participants' average across all trials and their scores on each trial.

Mean square for the subject by trial interaction: The average squared deviation between each individual's change in scores across trials and the average change in scores across trials.

Time, trial: Each time for which data were collected on the dependent variable.

Within-subject variance: Differences within each individual case on scores on the dependent variable across trials.

Within-subjects design: A repeated-measures ANOVA design in which intra-individual changes across trials were tested. This technique allows the researcher to test whether, on average, individuals scored differently at one time than another.

$MS_{s \times T}$: Symbol for the mean square for the interaction of subject by trial.

MS_T: Symbol for the mean square for the differences between the trials.

RECOMMENDED READING

Glass, G. V., & Hopkins, K. D. (1996). *Statistical methods in education and psychology* (3rd ed.). Boston: Allyn & Bacon.

Marascuilo, L. A., & Serlin, R. C. (1988). *Statistical methods for the social and behavioral sciences.* New York: Freeman.

CHAPTER 12

REGRESSION

In Chapter 7, the concept of correlation was introduced. Correlation involves a measure of the degree to which two variables are related to each other. A closely related concept, coefficient of determination, was also introduced in that chapter. This statistic provides a measure of the strength of the association between two variables in terms of percentage of variance explained. Both of these concepts are present in **regression.** In this chapter, the concepts of **simple linear regression** and **multiple regression** are introduced.

Regression is a very common statistic in the social sciences. One of the reasons it is such a popular technique is because it is so versatile. Regression, particularly multiple regression, allows researchers to examine how variables are related to each other, the strength of the relations, and relative predictive power of several independent variables on a dependent variable, and the unique contribution of one or more independent variables when controlling for one or more covariates. It is also possible to test for interactions in multiple regression. With all of the possible applications of multiple regression, it is clear that it is impossible to describe all of the functions of regression in this brief chapter. Therefore, the focus of this chapter is to provide an introduction to the concept and uses of regression, and to refer the reader to resources to examine for additional information.

Simple Versus Multiple Regression

The difference between simple and multiple regression is similar to the difference between one-way and factorial ANOVA. Like one-way ANOVA, simple regression analysis involves a single **independent,** or **predictor variable** and a single **dependent,** or **outcome variable.** This is the same number of variables used in a simple correlation analysis. The difference between a Pearson correlation coefficient and simple regression analysis is that whereas the correlation does not distinguish between independent and dependent variables, in a regression analysis there is always a designated predictor variable and a designated dependent variable. That is because the purpose of regression analysis is to make *predictions* about the value of the dependent variable given certain values of the predictor variable. This is a simple extension of correlation analyses. If I am interested in the relationship between height and weight, for example, I could use simple regression analysis to answer this question: If I know a man's height, what would I *predict* his weight to be? Of course, the accuracy of my prediction will only be as good as my correlation will allow, with stronger correlations leading to more accurate predictions. Therefore, simple linear regression is not really a more powerful tool than simple correlation analysis. But it does give me another way of conceptualizing the relation between two variables, a point I elaborate on shortly.

The real power of regression analysis can be found in multiple regression. Like factorial ANOVA, multiple regression involves models that have two or more predictor variables and a single dependent variable. For example, suppose that, again, I am interested in predicting how much a person weighs (i.e., weight is the dependent variable). Now, suppose that in addition to height, I know how many minutes of exercise the person gets per day, and how many calories a day he consumes. Now I've got three predictor variables (height, exercise, and calories consumed) to help me make an educated guess about the person's weight. Multiple regression analysis allows me to see, among other things, (a) how much these three predictor variables, as a group, are related to weight, (b) the strength of the relationship between each predictor variable and the dependent variable *while controlling for the other predictor variables in the model,* (c) the *relative* strength of each predictor variable, and (d) whether there are interaction effects between the predictor variables. As you can see, multiple regression is a particularly versatile and powerful statistical technique.

Variables Used in Regression

As with correlation analysis, in regression the dependent and independent variables need to be measured on an interval or ratio scale. **Dichotomous** (i.e., categorical variables with two categories) predictor variables can also be used. There is a special form of regression analysis, logit regression, that allows us to examine dichotomous dependent variables, but this type of regression is beyond the scope of this book. In this chapter, we limit our consideration of regression to those types that involve a continuous dependent variable and either continuous or dichotomous predictor variables.

REGRESSION IN DEPTH

Regression, particularly simple linear regression, is a statistical technique that is very closely related to correlations (discussed in Chapter 7). In fact, when examining the relationship between two continuous (i.e., measured on an interval or ratio scale) variables, either a correlation coefficient or a regression equation can be used. Indeed, the Pearson correlation coefficient is nothing more than a simple linear regression coefficient that has been standardized. The benefits of conducting a regression analysis rather than a correlation analysis are (a) regression analysis yields more information, particularly when conducted with one of the common statistical software packages, and (b) the regression equation allows us to think about the relation between the two variables of interest in a more intuitive way. Whereas the correlation coefficient provides us with a single number (e.g., $r = .40$), which we can then try to interpret, the regression analysis yields a formula for calculating the **predicted value** of one variable when we know the actual value of the second variable. Here's how it works.

The key to understanding regression is to understand the formula for the regression equation. So I begin by presenting the most simple form of the regression equation, describe how it works, and then move on to more complicated forms of the equation. In Table 12.1, the regression equation used to find the predicted value of Y, along with definitions of the components, are presented.

TABLE 12.1 The regression equation.

$$\hat{Y} = b\mathrm{X} + a$$

where $\hat{Y}$ is the predicted value of the Y variable
  b is the unstandardized regression coefficient, or the slope
  a is intercept (i.e., the point where the regression line intercepts the Y axis)

In simple linear regression, we begin with the assumption that the two variables are **linearly** related. In other words, if the two variables are actually related to each other, we assume that every time there is an increase of a given size in value on the X variable (called the **predictor** or **independent** variable), there is a corresponding increase (if there is a positive correlation) or decrease (if there is a negative correlation) *of a given size* in the Y variable (called the **dependent**, or **outcome**, or **criterion** variable). In other words, if the value of X increases from a value of 1 to a value of 2, and Y increases by 2 points, then when X increases from 2 to 3, we would predict that the value of Y would increase another 2 points.

To illustrate this point, let's consider the following set of data. Suppose I wanted to know whether there was a relationship between the amount of education people have and their monthly income. Education level is measured in years, beginning with kindergarten and extending through graduate school. Income is measured in thousands of dollars. Suppose that I randomly select a sample of 10 adults and measure their level of education and their monthly income, getting the data provided in Table 12.2.

When we look at these data, we can see that, in general, monthly income increases as the level of education increases. This is a general, rather than an absolute, trend because in some cases a person with more years of education makes less money per month than someone with less education (e.g., Case 10 and Case 9, Case 6 and Case 5). So although not every person with more education makes more money, *on average* more years of education are associated with higher monthly incomes. The correlation coefficient that describes the relation of these two variables is $r = .83$, which is a very strong, positive correlation (see Chapter 7 for a more detailed discussion of correlation coefficients).

TABLE 12.2 Income and education level data.

	Education Level (X)	Monthly Income (Y) (in thousands)
Case 1	6 years	1
Case 2	8 years	1.5
Case 3	11 years	1
Case 4	12 years	2
Case 5	12 years	4
Case 6	13 years	2.5
Case 7	14 years	5
Case 8	16 years	6
Case 9	16 years	10
Case 10	21 years	8
Mean	12.9	4.1
Standard Deviation	4.25	3.12
Correlation Coefficient	.83	

If we were to plot these data on a simple graph, we would produce a **scatterplot**, such as the one provided in Figure 12.1. In this scatterplot, there are 10 data points, one for each case in the study. Note that each data point marks the spot where education level (the X variable) and monthly income (the Y variable) meet for each case. For example, the point that has a value of 10 on the Y axis (income) and 16 on the X axis (education level) is the data point for the 10th case in our sample. These 10 data points in our scatterplot reveal a fairly distinct trend. Notice that the points rise somewhat uniformly from the lower left corner of the graph to the upper right corner. This shape is a clear indicator of the positive relationship (i.e., correlation) between education level and income. If there had been a perfect correlation between these two variables (i.e., $r = 1.0$), the data points would be aligned in a perfectly straight line, rising from lower left to upper right on the graph. If the relationship between these two variables were weaker (e.g., $r = .30$), the data points would be more widely scattered, making the lower-left to upper-right trend much less clear.

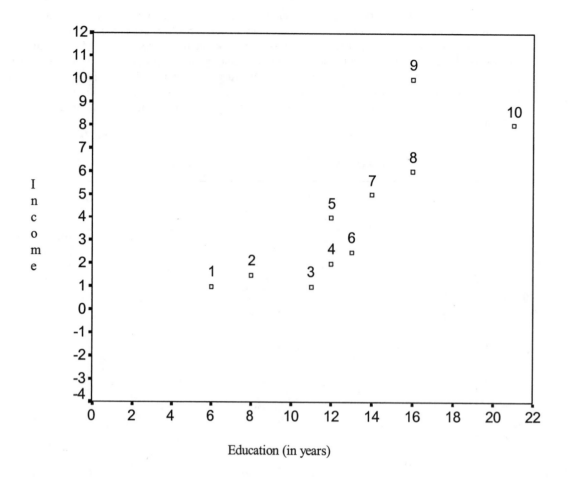

FIGURE 12.1 Scatterplot for education and income.

With the data provided in Table 12.2, we can calculate all of the pieces of the regression equation. The regression equation allows us to do two things. First, it lets us find predicted values for the Y variable for any given value of the X variable. In other words, we can predict a person's monthly income if we know how many years of education he or she has. Second, the regression equation allows us to produce the **regression line**. The regression line is the basis for linear regression and can help us understand how regression works.

There are a number of different types of regression formulas, but the most commonly used is called **ordinary least squares regression**, or **OLS.** OLS is based on an idea that we have seen before: the *sum of squares* (see Chapters 2 and 9). If you wanted to, you could draw a number of straight lines that bisect the data points presented in the scatterplot in Figure 12.1. For example, you could draw a horizontal line that extends out from the number 5 on the Y axis. Similarly, you could draw a straight line that extends down from the number 10 on the Y axis to the number 25 on the X axis. No matter how you decided to draw your straight line, notice that at least some of the data points in the scatterplot will not fall exactly *on* the line. Some or all will fall above the line, some may fall directly on the line, and some or all will fall below the line. Any data point that does not fall directly on the line will have a certain amount of distance between the point and the line. Now if you were to calculate the distance between the data point and the line you have drawn, and then square that distance, you would have a *squared deviation* for that point. If you calculated the squared deviation for each data point that did not fall on the line, and added all of these squared deviations together, you would end up with the *sum of squared deviations, or sum of squares.*

Now here is the key: The sum of the squared deviations, or sum of squares, will differ depending on where you draw your line. In any scatterplot, there is only *one* line the produces the *smallest* sum of squares. This line is known as the line of *least squares*, and this is the regression line. So, the reason this type of regression is called ordinary *least squares* regression is because in this type of regression, the regression line represents the straight line that produces the *smallest* sum of squared deviations from the line. This regression line represents the *predicted* values of Y at any given value of X. Of course, when we predict a value of Y for a given value of X, our prediction may be off. This error in prediction is represented by the distance between the regression line and the actual data point(s) in the scatterplot. To illustrate how this works, we first need to calculate the properties of the regression line (i.e., its slope and intercept). Then, we draw this regression line into the scatterplot, and you can see how well it "fits" the data (i.e., how close the data points fall to the regression line).

If you take a look at the formula for the regression equation in Table 12.1, you will that there are four components: (a) $\hat{Y}$ is the predicted value of the Y variable, (b) **b** is the unstandardized **regression coefficient**, and is also the **slope** of the regression line, (c) **X** is the value of the X variable, and (d) **a** is the value of the **intercept** (i.e., where the regression line crosses the Y axis). Because Y is the value produced by **regression equation**, let's save that one for last. And because X is just a given value on the X variable, there is not really anything to work out with that one. So let's take a closer look at a and b.

We cannot calculate the intercept before we know the slope of the regression line, so let's begin there. The formula for calculating the regression coefficient is

$$b = r * \frac{s_y}{s_x}$$

where b is the regression coefficient,
r is the correlation between the X and Y variables,
s_y is the standard deviation of the Y variable,
s_x is the standard deviation of the X variable.

Looking at the data in Table 12.2, we can see that $r = .83$, $s_y = 3.12$, $s_x = 4.25$. When we plug these numbers into the formula, we get the following:

$$b = .83 * \frac{3.12}{4.25}$$

$$b = (.83) * (.73)$$

$$b = .61$$

Notice that the regression coefficient is simply the correlation coefficient times the ratio of the standard deviations for the two variables involved. When we multiply the correlation coefficient by this ratio of standard deviations, we are roughly transforming the correlation coefficient into the scales of measurement used for the two variables. Notice that there is a smaller range, or less variety, of scores on our Y variable that there is on our X variable in this example. This is reflected in the ratio of standard deviations used to calculate b.

Now that we've got our b, we can calculate our intercept, a. The formula for a is as follows:

$$a = \overline{Y} - bX$$

where $\overline{Y}$ is the average value of Y,
X is the average value of X,
and b is the regression coefficient.

When we plug in the values from Table 12.2, we find that

$$a = 4.1 - (.61)(12.9)$$

$$a = 4.1 - 7.87$$

$$a = -3.77.$$

This value of a indicates that the intercept for the regression line is -3.77. In other words, the regression line crosses the Y axis at a value of -3.77. In still other words, this intercept tells us that when $X = 0$, we would predict the value of Y to be -3.77. Of course, in the real world, it is not possible to have a monthly income of negative 3.77 thousand dollars. Such unrealistic values remind us that we are dealing with *predicted* values of Y. Given our data, if a person has absolutely no formal education, we would *predict* that person to make a negative amount of money.

Now we can start to fill out our regression equation. The original formula

$$\hat{Y} = a + bX$$

now reads

$$\hat{Y} = -3.77 + .61X.$$

It is important to remember that when we use the regression equation to find predicted values of Y for different values of X, we are not calculating the *actual* value of Y. We are only making *predictions* about the value of Y. Whenever we make predictions, we will sometimes be incorrect. Therefore, there is bound to be some **error** (*e*) in our predictions about the values of Y at given values of X. The stronger the relationship (i.e., correlation) between my X and Y variables, the less error there will be in my predictions. The error is the difference between the actual, or observed, value of Y and the predicted value of Y. Because the predicted value of Y is simply $a + bX$, we can express the formula for the error in two ways:

$$e = Y - \hat{Y}$$

$$e = Y - a + bX$$

So rather than a single regression equation, there are actually two. One of them, the one presented in Table 12.1, is for the *predicted* value of Y ($\hat{Y}$). The other one is for the actual, or *observed* value of Y. This equation takes into account the errors in our predictions, and is written as $Y = bX + a + e$.

Now that we've got our regression equation, we can put it to use. First, let's wrap words around it, so that we can make sure we understand what it tells us. Our regression coefficient tells us that "For every unit of increase in X, there is a corresponding predicted increase of .61 units in Y." Applying this to our variables, we can say that "For every additional year of education, we would *predict* an increase of .61 ($1,000), or $610, in monthly income." We know that the predicted value of Y will *increase* when X increases, and vice-versa, because the regression coefficient is *positive*. Had it been negative, we would predict a decrease in Y when X increases.

Next, let's use our regression equation to find predicted values of Y at given values of X. For example, what would we predict the monthly income to be for a person with 9 years of formal education? To answer this question, we plug in the value of 9 for the X variable and solve the equation:

$$\hat{Y} = -3.77 + .61(9)$$

$$\hat{Y} = -3.77 + 5.59$$

$$\hat{Y} = 1.82$$

So we would predict that a person with 9 years of education would make $1,820 per month, plus or minus our error in prediction (*e*).

Finally, we can use our regression equation to compute our regression line. We already know, from the value of the intercept, that our regression line will cross the *Y* axis at a value of -3.77. To draw a straight line, all we need to do is calculate one additional point. To make sure we include all of the points in our scatterplot, let's just calculate a predicted value of *Y* for a person with 25 years of education.

$$\hat{Y} = -3.77 + .61(25)$$

$$\hat{Y} = -3.77 + 15.25$$

$$\hat{Y} = 11.48$$

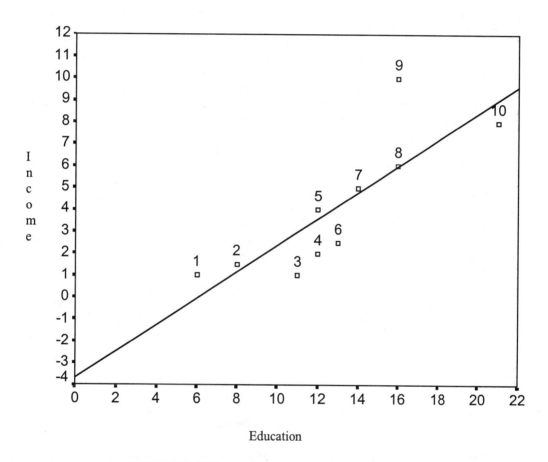

FIGURE 12.2 Scatterplot with regression line.

If we were to draw a regression line through our scatterplot, using the two points we found from our intercept and the predicted *Y* value, we would get something like the line presented in Figure 12.2.

With the regression line added to the scatterplot, some of the concepts mentioned earlier in this chapter may be easier to understand. First, notice that our regression line does not accurately predict the actual *Y* values for any of our cases except for Case 8. That data point is precisely on the regression line. For each of the other nine cases, there is some amount of error present in the prediction. In some cases, the amount of error is very little (e.g., Case 7), whereas in others the amount of error is quite large (e.g., Case 9). These errors in prediction are known as **residuals**. In some cases, our predicted value was greater than our **observed value** (e.g., Cases 1, 2, 5, 7, and 9). For these cases, we have **overpredicted** their income based on their level of education. Such overpredictions produce *negative* residuals (because the residual = observed scores - predicted score). For other cases (Cases 3, 4, 6, and 10) we **underpredicted** the *Y* value, creating *positive* residuals. Second, notice the distance between each case and the line. When we square each of these distances, and then add them all together, we get the sum of squares. Third, notice that the regression line marks the line where the sum of the squared distances is smallest. To test this, try drawing some other lines and noting the way it increases the overall amount of error in prediction. Finally, notice where the regression line crosses the *Y* axis (the intercept) and the how much higher up the *Y* axis the regression line goes for each increase of one unit value in *X* (the slope). The slope and the intercept will correspond with the values that we found for *b* and *a*, respectively.

MULTIPLE REGRESSION

Now that we've discussed the elements of simple linear regression, let's move on to a consideration of **multiple regression**. Despite the impressive qualities of simple linear regression, the plain truth is that when we only have two variables, simple linear regression does not provide much more information than would a simple correlation coefficient. Because of this, you rarely see a simple linear regression with two variables reported in a published study. But multiple regression is a whole different story. Multiple regression is a very powerful statistic that can be used to provide a staggering array of useful information. At this point, it may be worth reminding you that in a short book like this, we only scratch the surface of what multiple regression can do and how it works. The interested reader should refer to one or all of the references listed at the end of this chapter to find more information on this powerful technique.

To illustrate some of the benefits of multiple regression, let's add a second *predictor* variable to our example. So far, using the data from Table 12.2, we have examined the relationship between education level and income. In this example, education level has been used as our *predictor* or *independent* variable and income has been used as our *dependent* or *outcome* variable. We found that, on average, in our sample one's monthly salary is predicted to increase by $610 for every additional year of schooling the individual has received. But there was some error in our predictions, indicating that there are other variables that predict how much money one makes. One such predictor may be the length of time one has been out of school. Because people tend to make more money the longer they have been in the workforce, it stands to reason that those adults in our sample who finished school a long time ago may be making more than those who finished school more recently. Although Case 4 and Case 5 each had 12 years of schooling, Case 5 makes more money than Case 4. Perhaps this is due to Case 5 being in the workforce longer than Case 4.

When we add this second predictor variable to the model, we get the following regression equation:

$$\hat{Y} = a + bX_1 + bX_2$$

where $\hat{Y}$ is the predicted value of the dependent variable,
X_1 is the value of the first predictor variable,
and X_2 is the value of the second predictor variable.

This regression equation with two predictor variables will allow me to examine a number of different questions. First, I can see whether my two predictor variables, combined, are significantly related to, or predictive of, my dependent variable, and how much of the variance my predictor variables explain in my dependent variable. Second, I can test whether each of my predictor variables is significantly related to my dependent variable *when controlling for the other predictor variable*. When I say "controlling for the other predictor variable" I mean that I can examine whether a predictor variable is related to the dependent variable after I partial out, or take away, the portion of the variance in my dependent variable that has already been accounted for by my other independent variable. Third, I can see which of my two predictor variables is the stronger predictor of my dependent variable. Fourth, I can test whether one predictor variable is related to my dependent variable after controlling for the other predictor variable, thus conducting a sort of ANCOVA (see Chapter 10 for a discussion of ANCOVA). There are many other things I can do with multiple regression, but I will limit my discussion to these four. (Note: Perhaps the most important omission from this discussion is the ability to examine statistical interactions between the predictor variables in multiple regression. This procedure is a bit too complex for this book, but excellent discussions of the technique can be found in Aiken & West, 1991, or Jaccard, Turrisi & Wan, 1990).

Suppose that for my 10 cases in my sample I also measured the number of years that they have been in the workforce, and get the data presented in Table 12.3.

TABLE 12.3 Income and education level data.

	Education Level (X₁)	Years Working (X₂)	Monthly Income (Y) (in thousands)
Case 1	6 years	10	1
Case 2	8 years	14	1.5
Case 3	11 years	8	1
Case 4	12 years	7	2
Case 5	12 years	20	4
Case 6	13 years	15	2.5
Case 7	14 years	17	5
Case 8	16 years	22	6
Case 9	16 years	30	10
Case 10	21 years	10	8
Mean	12.9	15	4.1
Standard Deviation	4.25	7.20	3.12
Correlation with Income	$r = .83$	$r = .70$	

The data presented in Table 12.3 reveal that both years of education and years in the workforce are positively correlated with monthly income. But how much of the variance in income can these two predictor variables explain *together*? Will years of education still predict income *when we control for the effects of years in the workforce*? In other words, after I partial out the portion of the variance in income that is accounted for by years in the workforce, will years of education still be able to help us predict income? Which of these two independent variables will be the stronger predictor of income? And will each make a unique contribution in explaining variance in income?

To answer these questions, I use the SPSS statistical software package to analyze my data. (Note: With only 10 cases in my sample, it is not wise to run a multiple regression. I am doing so for illustration purposes only. When conducting multiple regression analyses, you should have at least 30 cases plus 10 cases for each predictor variable in the model.) I begin by computing Pearson correlation coefficients for all three of the variables in the model. The results are presented in Table 12.4.

TABLE 12.4   Correlations among variables in regression model.

	Years of Education	*Years in Workforce*	*Monthly Income*
Years of Education	1.000		
Years in Workforce	.310	1.000	
Monthly Income	.826	.695	1.00

These data reveal that both level of education and years in the workforce are both correlated with monthly income ($r = .826$ and $r = .695$ for education and workforce with income, respectively). In Table 12.4 we can also see that there is a small-to-moderate correlation between our two predictors, years of education and years in the workforce ($r = .310$). Because this correlation is fairly weak, we can infer that both of these independent variables may predict education level.

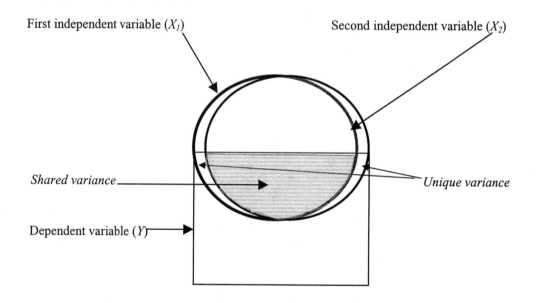

FIGURE 12.3   Shared variance in multiple regression.

Remember that in a multiple regression, we've got multiple predictor variables trying to explain variance in the dependent variable. For a predictor variable to explain variance in a dependent variable, it must be *related* to the dependent variable (see Chapter 7 and the discussion on the coefficient of determination). In our current example, both of our predictor variables are strongly correlated with our dependent variable, so this condition is met. In addition, for each of our predictor variables to explain a *unique*, or *independent* portion of the variance in the dependent variable, our two predictor variables cannot be too strongly related to *each other*. If our two predictor variables are strongly correlated with each other, then there is not going to be very much unexplained variance *in my predictor variables* left over to explain variance in the dependent variable. For example, suppose that the correlation between scores on a reading test were strongly correlated with scores on a writing test ($r = .90$). Now suppose that I wanted to use reading and writing test scores to predict students' grades in the English class. Because reading and writing test scores are so highly correlated with each other, I will probably not explain any more of the variance in English class grades using both predictor variables than if I had just used one or the other. In other words, once I use reading test scores to predict English class grades, adding writing test scores to my regression model will probably not explain any more of the variance in my dependent variable, because reading and writing test scores are so closely related to each other. This concept is represented graphically in Figure 12.3. The shaded area represents shared variance. Notice that the shaded area in the two predictor variables is so large, it is virtually impossible for any of the unshaded areas in each predictor variable to overlap with the dependent variable. These unshaded areas represent the **unique variance** explaining power of each predictor. You can see that when these unique portions of the predictor variables are small, it is difficult for each predictor to explain a unique portion of the variance in the dependent variable. Strong correlations among predictor variables is called **multicollinearity** and can cause problems in multiple regression analysis because it can make it difficult to identify the unique relation between each predictor variable and the dependent variable.

TABLE 12.5 Sample multiple regression results predicting monthly income.

		Variance Explained		
	R	*R Square*	*Adjusted R Square*	*Std. Error of the Estimate*
	.946	.896	.866	1.1405

	ANOVA Results				
	Sum of Squares	*df*	*Mean Square*	*F Value*	*p Value*
Regression	78.295	2	39.147	30.095	.000
Residual	9.105	7	1.301		
Total	87.400	9			

	Regression Coefficients				
	Unstandardized Coefficients		*Standardized Coefficients*		
	B	*Std. Error*	*Beta*	*t Value*	*p Value*
Intercept	-5.504	1.298		-4.241	.004
Years Education	.495	.094	.676	5.270	.001
Years Work	.210	.056	.485	3.783	.007

Returning to our example of using education level and years in the workforce to predict monthly income, when I conduct the regression analysis using SPSS, I get the results presented in Table 12.5. There are a variety of results produced with a multiple regression model. These results have been organized into three sections in Table 12.5. I have labeled the first section "Variance Explained." Here, we can see that we get an "R" value of .946. This is the **multiple correlation coefficient (R)**, and it provides a measure of the correlation between the two predictors *combined* and the dependent variable. It is also the correlation between the observed value of Y and the predicted value of Y ($\hat{Y}$). So together, years of education and years in the workforce have a very strong correlation with monthly income. Next, we get an "R Square" value (symbolized R^2). This is essentially the coefficient of determination (see Chapter 7) for my combined predictor variables and the dependent variables, and it provides us with a percentage of variance explained. So years of education and years in the workforce, combined, explain 89.6% of the variance in monthly income. When you consider that this leaves only about 10% of the variance in monthly income unexplained, you can see that this is a very large amount of variance explained. The R^2 statistic is the measure of effect size used in multiple regression. Because it is a measure of variance explained (like r^2 in correlation and eta^2 in ANOVA), it provides a handy way of assessing the practical significance of the relation of the predictors to the dependent variable. In this example, the effect size is large suggesting practical significance as well as statistical significance. The "Adjusted R Square" accounts for some of the error associated with multiple predictor variables by taking the number of predictor variables and the sample size into account, and thereby adjusts the R^2 value down a little bit. Finally, there is a standard error for the R and R^2 value (see Chapter 5 for a discussion of standard errors).

Moving down the table to the "ANOVA Results" section we get some statistics that help us determine whether our overall regression model is statistically significant. This section simply tells us whether our two predictor variables, combined, are able to explain a statistically significant portion of the variance in our dependent variable. The F value of 30.095, with a corresponding p value of .000, reveals that our regression model is statistically significant. In other words, the relationship between years of education and years in the workforce combined (our predictor variables) and monthly income (our dependent variable) is statistically significant (i.e., greater than zero). Notice that these ANOVA statistics are quite similar to those presented in Chapter 10, gender and GPA to predict feelings of self-efficacy among high school students. The sum of squares model in Table 10.2 corresponds to the sum of squares regression in Table 12.5. In both cases, we have sums of squares associated with the *combined predictors*, or the overall model.

Similarly, the sum of squares error in Table 10.2 is analogous to the sum of squares residual in Table 12.5. That is because residuals are simply another form of error. Just as the overall F value in Table 10.2 is produced by dividing the mean squares for the model by the mean squares error, the overall F value produced in Table 12.5 is produced by dividing the means squares regression by the mean squares residual. In both cases, we get an F value, and a corresponding significance test, which indicates whether, overall, our predictors are significantly related to our dependent variable.

Finally, in the third section of Table 12.5, we get to the most interesting part of the table. Here we see our intercept and the regression coefficients for each predictor variable. These are the pieces of the regression equation. We can use these statistics to create the regression equation:

$$\hat{Y} = -5.504 + .495X_1 + .210X_2$$

where $\hat{Y}$ is the predicted value of Y,
X_1 is the value of the years of education variable,
and X_2 is the value of the years in the workforce variable.

The unstandardized regression coefficients can be found in the column labeled "B." Because years of education and years in the workforce are variables with different standard deviations, it is difficult to compare the size of the unstandardized regression coefficients. The variables are simply measured on different scales, making comparisons difficult. However, in the column labeled "Beta", the **standardized regression coefficients** are presented. These regression coefficients have been standardized, thereby converting the unstandardized coefficients into coefficients with the same scale of measurement (z scores; see Chapter 4 for a discussion of standardization). Here we can see that the two predictors are fairly close in their strength of relation to the dependent variable, but years of education is a bit stronger than years of work. In the next two columns, labeled "t value" and "p value" we get measures that allow us to determine whether each predictor variable is statistically significantly related to the dependent variable. Recall that earlier, in the ANOVA section of the table, we saw that the two predictor variables *combined* were significantly related to the dependent variable. Now we can use t tests to see if the slope for *each* predictor variable is significantly different from zero. The p values associated with each predictor variable are well smaller than .05, indicating that each of my independent variables is a significant predictor of my dependent variable. So both years of education and years in the workforce are statistically significant predictors of monthly income.

It is important to note that in this last section of Table 12.5, each regression coefficient shows the strength of the relationship between the predictor variable and the outcome variable *while controlling for the other predictor variable*. Recall that in the simple regression model with one predictor variable, I found that there was a relationship between years of education and monthly income. One of my questions in the multiple regression model was whether this education-income link would remain statistically significant when controlling for, or partialing out, the effects of years in the workforce. As the results presented in Table 12.5 indicate, even when controlling for the effects of years in the workforce, years of education is still a statistically significant predictor of monthly income. Similarly, when controlling for years of education, years in the workforce predicts monthly income as well.

As you can see, multiple regression provides a wealth of information about the relations between predictor variables and dependent variables. Amazingly, in our previous example, we have just scratched the surface of all that can be done with multiple regression analysis. Therefore, I strongly encourage you to read more about multiple regression using the references provided at the end of this chapter. I also want to caution you about how to interpret regression analyses, whether they be simple or multiple regressions. Despite the uses of such terms as *predictor* and *dependent* variables, it is important to remember that regression analysis is based on good old correlations. Just as correlations should not be mistaken for proof of causal relationships between variables, regression analyses cannot prove that one variable, or set of variables, causes variation in another variable. Regression analyses can reveal how sets of variables are *related* to each other, but cannot prove causal relations among variables.

EXAMPLE: PREDICTING THE USE OF SELF-HANDICAPPING STRATEGIES

Sometimes students engage in behaviors that actually undermine their chances of succeeding academically. For example, they may procrastinate rather than study for an upcoming test, or they may spend time with their friends when they should be doing their homework. These behaviors are called "self-handicapping" because they actually inhibit students' chances of succeeding. One reason that students may engage in such behaviors is to provide an explanation for their poor academic performance, should it occur. If students fears that they may perform poorly on an academic task, they may not want others to think that the reason for this poor performance is that they lack ability, or intelligence. So some students strategically engage in self-handicapping to provide an alternative explanation for the poor performance. That is why these behaviors are called self-handicapping *strategies*.

Because self-handicapping strategies can undermine academic achievement, and may be a sign of academic withdrawal on the part of students, it is important to understand the factors that are associated with the use of these strategies. Self-handicapping represents a concern with not looking academically unable, even if that means perhaps sacrificing performance. Therefore, engaging in self-handicapping behaviors may be related to students' goals of avoiding appearing academically unable to others. In addition, because self-handicapping may be provoked by performance situations in which students expect to fail, perhaps it occurs more commonly among lower achieving students, who have a history of poor academic performance. Moreover, it is reasonable to suspect that when students lack confidence in their academic abilities, they will be more likely to use self-handicapping strategies. Finally, there may be gender differences in how concerned high school students are with looking academically unable to others. Therefore, I conducted a multiple regression analysis to examine whether avoidance goals, self-efficacy, gender, and GPA, as a group and individually, predicted the use of self-handicapping strategies.

My colleague, Carol Giancarlo, and I recently collected data from 464 high school students in which we used surveys to measure their self-reported use of self-handicapping strategies. In addition, the survey contained questions about their desire to avoid looking academic unable (called "avoidance goals"), and their confidence in their ability to perform academically (called "self-efficacy"). We also collected information about the students' gender (i.e., whether they were boys or girls) and their overall GPA in high school. Self-handicapping, avoidance goals, and self-efficacy were all measured using a 1-5 scale. Low scores indicated that students did not believe the items were true for them (i.e., did not use self-handicapping strategies, were not confident in their abilities, were not concerned with trying to avoid looking academically unable) whereas high scores indicated the opposite. Gender was "dummy" coded (boys = 1, girls = 0), and GPA was measured using a scale from 0 to 4.0 (0 = F, 4.0 = A average).

Once again, I used SPSS to analyze my data. The results of this multiple regression analysis are presented in Table 12.6. In the first section of the table, "Variance Explained," we can see there is an R value of .347, and an R^2 value of .12. These statistics tell us that the four predictor variables, combined, have a moderate correlation with self-handicapping (multiple R = .347), and explain 12% of the variance in handicapping. This R^2 value is reduced to .113 when adjusted for the error associated with multiple predictor variables. As I move down to the second section of the table, "ANOVA Results," I see that I have an F value of 15.686 and a corresponding p value of .000. These results tell me that, as a group, my four predictor variables explain a statistically significant portion of the variance in self-handicapping. In other words, my overall regression model is statistically significant.

TABLE 12.6 Multiple regression results: predicting self-handicapping.

		Variance Explained		
	R	R Square	Adjusted R Square	Std. Error of the Estimate
	.347	.120	.113	.9005

		ANOVA Results			
	Sum of Squares	df	Mean Square	F Value	p Value
Regression	50.877	4	12.719	15.686	.000
Residual	372.182	459	.811		
Total	423.059	463			

			Regression Coefficients		
	Unstandardized Coefficients		Standardized Coefficients		
	B	Std. Error	Beta	t Value	p Value
Intercept	3.630	.264		13.775	.000
Avoidance Goals	.132	.045	.130	2.943	.003
Grades (GPA)	-.254	.054	-.209	-4.690	.000
Gender	.105	.085	.055	1.234	.218
Self Efficacy	-.232	.052	-.198	-4.425	.000

In the last section of the table, I find my unstandardized regression coefficients (column labeled "B") for each predictor variable in the model, as well as my intercept. These tell me that GPA and self-efficacy are negatively related to self-handicapping whereas gender and avoidance goals are positively related to self-handicapping. Scanning toward the right side of the table, I find the standardized regression coefficients (column labeled "Beta"). These coefficients, which are all converted to the same, standardized scale, reveal that GPA and self-efficacy appear to be more strongly related to self-handicapping than are avoidance goals and, in particular, gender. Continuing to scan toward the right side of the table, I find my *t* values and *p* values for each coefficient. These tell me which of my independent variables are statistically significant predictors of self-handicapping. The *p* values tell me that all of my independent variables except for gender are significant predictors of handicapping.

So what can we make of these results? First, my predictors explain a significant percentage of the variance in self-handicapping, although not a particularly large percentage (about 11%). Second, as we might expect, students with higher GPAs report engaging in less self-handicapping behavior than students with lower GPAs. Third, students with more confidence in their academic abilities engage in less self-handicapping than do students with less confidence in their abilities. Fourth, students who are concerned with not looking academically unable in school are more likely to use self-handicapping strategies than are students without this concern. Finally, boys and girls do not differ significantly in their reported use of self-handicapping strategies. Although boys scored slightly higher than girls on the handicapping items (we know this because the regression coefficient was positive, and the gender variable was coded boys = 1, girls = 0), this difference was not statistically significant.

It is important to remember that the results for each independent variable are reported while controlling for the effects of the other independent variables. So the statistically significant relationship between self-efficacy and self-handicapping exists even when we control for the effects of GPA and avoidance goals. This is important, because one may be tempted to argue that the relationship between confidence and self-handicapping is merely a by-product of academic achievement. Those who perform better in school *should* be more confident in their abilities, and therefore *should* engage in less self-handicapping. What the results of this multiple regression reveal is that there is a statistically significant relationship between self-efficacy and self-handicapping *even after controlling for the effects of academic performance.* Confidence is associated with less self-handicapping *regardless* of one's level of academic achievement. Similarly, when students are concerned with not looking dumb in school (avoidance goals), *regardless* of their actual level of achievement (GPA), they are more likely to engage in self-handicapping behavior. The ability to examine both the combined and independent relations among predictor variables and a dependent variable is the true value of multiple regression analysis.

WRAPPING UP

The overlap between correlations (Chapter 7) and regression are plain. In fact, simple linear regression provides a statistic, the regression coefficient, that is simply the unstandardized version of the Pearson correlation coefficient. What may be less clear, but equally important, is that regression is also a close relative of ANOVA. As you saw in the discussion of Table 12.6, regression *is* a form of analysis of variance. Once again, we are interested in dividing up the variance of a dependent variable and explaining it with our independent variables. The major difference between ANOVA and regression generally involves the types of variables that are analyzed, with ANOVA using categorical independent variables and regression using continuous independent variables. As you learn more about regression on your own, you will learn that even this simple distinction is a false one, as categorical independent variables can be analyzed in regression.

The purpose of this book was to provide plain English explanations of the most commonly used statistical techniques. Because this book is short, and because there is only so much about statistics that can be explained in plain English, I hope that you will consider this the beginning of your journey into the statistical world rather than the end. Although sometimes intimidating and daunting, the world of statistics is also rewarding and worth the effort. Whether we like it or not, all of our lives are touched and, at times, strongly affected by statistics. It is important that we make the effort to understand how statistics work and what they mean. If you have made it to the end of this book, you have already made substantial strides toward achieving that understanding. I'm sure that with continued effort, you will be able to take advantage of the many insights that an understanding of statistics can provide. Enjoy the journey.

GLOSSARY OF TERMS AND SYMBOLS FOR CHAPTER 12

Dependent, outcome, criterion variable: Different terms for the dependent variable.

Error: Amount of difference between the predicted value and the observed value of the dependent variable. It is also the amount of unexplained variance in the dependent variable.

Independent, predictor variable: Different terms for the independent variable.

Intercept: Point at which the regression line intersects the Y axis. Also, the value of Y when $X = 0$.

Multicollinearity: The degree of overlap among predictor variables in a multiple regression. High multicollinearity among predictor variables can cause difficulties finding unique relations among predictors and the dependent variable.

Multiple correlation coefficient: A statistic measuring the strength of the association between multiple independent variables, as a group, and the dependent variable.

Multiple regression: A regression model with more than one independent, or predictor, variable.

Observed value: The actual, measured value of the Y variable at a given value of X.

Ordinary least squares regression (OLS): A common form of regression that uses the smallest sum of squared deviations to generate the regression line.

Overpredicted: Observed values of Y at given values of X that are *below* the predicted values of Y (i.e., the values predicted by the regression equation).

Predicted values: Estimates of the value of Y at given values of X that are generated by the regression equation.

Regression coefficient: A measure of the relationship between each predictor variable and the dependent variable. In simple linear regression, this is also the slope of the regression line. In multiple regression, the various regression coefficients combine to create the slope of the regression line.

Regression equation: The components, including the regression coefficients, intercept, error term, and X and Y values that are used to generate predicted values for Y and the regression line.

Regression line: The line that can be drawn through a scatterplot of the data that best "fits" the data (i.e., minimizes the squared deviations between observed values and the regression line).

Residuals: Errors in prediction. The difference between observed and predicted values of Y.

Scatterplot: A graphic representation of the data along two dimensions (X and Y).

Simple linear regression: The regression model employed when there is a single dependent and a single independent variable.

Slope: The average amount of change in the Y variable for each one unit of change in the X variable.

Standardized regression coefficient: The regression coefficient converted into standardized values.

Underpredicted: Observed values of Y at given values of X that are above the predicted values of Y (i.e., the values predicted by the regression equation).

Unique variance: The proportion of variance in the dependent variable explained by an independent variable when controlling for all other independent variables in the model.

$\hat{Y}$ Symbol for the predicted value of Y, the dependent variable.
Y Symbol for the observed value of Y, the dependent variable.
b Symbol for the unstandardized regression coefficient.
a Symbol for the intercept.
e Symbol for the error term.
R Symbol for the multiple correlation coefficient.
R^2 Symbol for the percentage of variance explained by the regression model.

REFERENCES AND RECOMMENDED READING

Aiken, L. S., & West, S. G. (1991). *Multiple regression: Testing and interpreting interactions.* Newbury Park, CA: Sage.

Berry, W. D., & Feldman, S. (1985). *Multiple regression in practice.* Newbury Park, CA: Sage.

Cohen, J., & Cohen, P. (1975). *Applied multiple regression/correlation analysis for the behavioral sciences.* Hillsdale, NJ: Lawrence Erlbaum Associates.

Jaccard, J., Turrisi, R., & Wan, C. K. (1990). *Interaction effects in multiple regression.* Newbury Park, CA: Sage.

APPENDIXES

APPENDIX A   Area under the normal curve between μ and z and beyond z.

A	B	C	A	B	C	A	B	C
	Area between mean and	Area beyond		Area between mean and	Area beyond		Area between mean and	Area beyond
z	z	z	z	z	z	z	z	z
0.00	.0000	.5000	0.55	.2088	.2912	1.10	.3643	.1357
0.01	.0040	.4960	0.56	.2123	.2877	1.11	.3665	.1335
0.02	.0080	.4920	0.57	.2157	.2843	1.12	.3686	.1314
0.03	.0120	.4880	0.58	.2190	.2810	1.13	.3708	.1292
0.04	.0160	.4840	0.59	.2224	.2776	1.14	.3729	.1271
0.05	.0199	.4801	0.60	.2257	.2743	1.15	.3749	.1251
0.06	.0239	.4761	0.61	.2291	.2709	1.16	.3770	.1230
0.07	.0279	.4721	0.62	.2324	.2676	1.17	.3790	.1210
0.08	.0319	.4681	0.63	.2357	.2643	1.18	.3810	.1190
0.09	.0359	.4641	0.64	.2389	.2611	1.19	.3830	.1170
0.10	.0398	.4602	0.65	.2422	.2578	1.20	.3849	.1151
0.11	.0438	.4562	0.66	.2454	.2546	1.21	.3869	.1131
0.12	.0478	.4522	0.67	.2486	.2514	1.22	.3888	.1112
0.13	.0517	.4483	0.68	.2517	.2483	1.23	.3907	.1093
0.14	.0557	.4443	0.69	.2549	.2451	1.24	.3925	.1075
0.15	.0596	.4404	0.70	.2580	.2420	1.25	.3944	.1056
0.16	.0636	.4364	0.71	.2611	.2389	1.26	.3962	.1038
0.17	.0675	.4325	0.72	.2642	.2358	1.27	.3980	.1020
0.18	.0714	.4286	0.73	.2673	.2327	1.28	.3997	.1003
0.19	.0753	.4247	0.74	.2704	.2296	1.29	.4015	.0985
0.20	.0793	.4207	0.75	.2734	.2266	1.30	.4032	.0968
0.21	.0832	.4168	0.76	.2764	.2236	1.31	.4049	.0951
0.22	.0871	.4129	0.77	.2794	.2206	1.32	.4066	.0934
0.23	.0910	.4090	0.78	.2823	.2177	1.33	.4082	.0918
0.24	.0948	.4052	0.79	.2852	.2148	1.34	.4099	.0901
0.25	.0987	.4013	0.80	.2881	.2119	1.35	.4115	.0885
0.26	.1026	.3974	0.81	.2910	.2090	1.36	.4131	.0869
0.27	.1064	.3936	0.82	.2939	.2061	1.37	.4147	.0853
0.28	.1103	.3897	0.83	.2967	.2033	1.38	.4162	.0838
0.29	.1141	.3859	0.84	.2995	.2005	1.39	.4177	.0823
0.30	.1179	.3821	0.85	.3023	.1977	1.40	.4192	.0808
0.31	.1217	.3783	0.86	.3051	.1949	1.41	.4207	.0793
0.32	.1255	.3745	0.87	.3078	.1922	1.42	.4222	.0778
0.33	.1293	.3707	0.88	.3106	.1894	1.43	.4236	.0764
0.34	.1331	.3669	0.89	.3133	.1867	1.44	.4251	.0749
0.35	.1368	.3632	0.90	.3159	.1841	1.45	.4265	.0735
0.36	.1406	.3594	0.91	.3186	.1814	1.46	.4279	.0721
0.37	.1443	.3557	0.92	.3212	.1788	1.47	.4292	.0708
0.38	.1480	.3520	0.93	.3238	.1762	1.48	.4306	.0694
0.39	.1517	.3483	0.94	.3264	.1736	1.49	.4319	.0681
0.40	.1554	.3446	0.95	.3289	.1711	1.50	.4332	.0668
0.41	.1591	.3409	0.96	.3315	.1685	1.51	.4345	.0655
0.42	.1628	.3372	0.97	.3340	.1660	1.52	.4357	.0643
0.43	.1664	.3336	0.98	.3365	.1635	1.53	.4370	.0630
0.44	.1700	.3300	0.99	.3389	.1611	1.54	.4382	.0618
0.45	.1736	.3264	1.00	.3413	.1587	1.55	.4394	.0606
0.46	.1772	.3228	1.01	.3438	.1562	1.56	.4406	.0594
0.47	.1808	.3192	1.02	.3461	.1539	1.57	.4418	.0582
0.48	.1844	.3156	1.03	.3485	.1515	1.58	.4429	.0571
0.49	.1879	.3121	1.04	.3508	.1492	1.59	.4441	.0559
0.50	.1915	.3085	1.05	.3531	.1469	1.60	.4452	.0548
0.51	.1950	.3050	1.06	.3554	.1446	1.61	.4463	.0537
0.52	.1985	.3015	1.07	.3577	.1423	1.62	.4474	.0526
0.53	.2019	.2981	1.08	.3599	.1401	1.63	.4484	.0516
0.54	.2054	.2946	1.09	.3621	.1379	1.64	.4495	.0505

APPENDIX A (continued)

A	B Area between mean and	C Area beyond	A	B Area between mean and	C Area beyond	A	B Area between mean and	C Area beyond
z	z	z	z	z	z	z	z	z
1.65	.4505	.0495	2.22	.4868	.0132	2.79	.4974	.0026
1.66	.4515	.0485	2.23	.4871	.0129	2.80	.4974	.0026
1.67	.4525	.0475	2.24	.4875	.0125	2.81	.4975	.0025
1.68	.4535	.0465	2.25	.4878	.0122	2.82	.4976	.0024
1.69	.4545	.0455	2.26	.4881	.0119	2.83	.4977	.0023
1.70	.4554	.0446	2.27	.4884	.0116	2.84	.4977	.0023
1.71	.4564	.0436	2.28	.4887	.0113	2.85	.4978	.0022
1.72	.4573	.0427	2.29	.4890	.0110	2.86	.4979	.0021
1.73	.4582	.0418	2.30	.4893	.0107	2.87	.4979	.0021
1.74	.4591	.0409	2.31	.4896	.0104	2.88	.4980	.0020
1.75	.4599	.0401	2.32	.4898	.0102	2.89	.4981	.0019
1.76	.4608	.0392	2.33	.4901	.0099	2.90	.4981	.0019
1.77	.4616	.0384	2.34	.4904	.0096	2.91	.4982	.0018
1.78	.4625	.0375	2.35	.4906	.0094	2.92	.4982	.0018
1.79	.4633	.0367	2.36	.4909	.0091	2.93	.4983	.0017
1.80	.4641	.0359	2.37	.4911	.0089	2.94	.4984	.0016
1.81	.4649	.0351	2.38	.4913	.0087	2.95	.4984	.0016
1.82	.4656	.0344	2.39	.4916	.0084	2.96	.4985	.0015
1.83	.4664	.0336	2.40	.4918	.0082	2.97	.4985	.0015
1.84	.4671	.0329	2.41	.4920	.0080	2.98	.4986	.0014
1.85	.4678	.0322	2.42	.4922	.0078	2.99	.4986	.0014
1.86	.4686	.0314	2.43	.4925	.0075	3.00	.4987	.0013
1.87	.4693	.0307	2.44	.4927	.0073	3.01	.4987	.0013
1.88	.4699	.0301	2.45	.4929	.0071	3.02	.4987	.0013
1.89	.4706	.0294	2.46	.4931	.0069	3.03	.4988	.0012
1.90	.4713	.0287	2.47	.4932	.0068	3.04	.4988	.0012
1.91	.4719	.0281	2.48	.4934	.0066	3.05	.4989	.0011
1.92	.4726	.0274	2.49	.4936	.0064	3.06	.4989	.0011
1.93	.4732	.0268	2.50	.4938	.0062	3.07	.4989	.0011
1.94	.4738	.0262	2.51	.4940	.0060	3.08	.4990	.0010
1.95	.4744	.0256	2.52	.4941	.0059	3.09	.4990	.0010
1.96	.4750	.0250	2.53	.4943	.0057	3.10	.4990	.0010
1.97	.4756	.0244	2.54	.4945	.0055	3.11	.4991	.0009
1.98	.4761	.0239	2.55	.4946	.0054	3.12	.4991	.0009
1.99	.4767	.0233	2.56	.4948	.0052	3.13	.4991	.0009
2.00	.4772	.0228	2.57	.4949	.0051	3.14	.4992	.0008
2.01	.4778	.0222	2.58	.4951	.0049	3.15	.4992	.0008
2.02	.4783	.0217	2.59	.4952	.0048	3.16	.4992	.0008
2.03	.4788	.0212	2.60	.4953	.0047	3.17	.4992	.0008
2.04	.4793	.0207	2.61	.4955	.0045	3.18	.4993	.0007
2.05	.4798	.0202	2.62	.4956	.0044	3.19	.4993	.0007
2.06	.4803	.0197	2.63	.4957	.0043	3.20	.4993	.0007
2.07	.4808	.0192	2.64	.4959	.0041	3.21	.4993	.0007
2.08	.4812	.0188	2.65	.4960	.0040	3.22	.4994	.0006
2.09	.4817	.0183	2.66	.4961	.0039	3.23	.4994	.0006
2.10	.4821	.0179	2.67	.4962	.0038	3.24	.4994	.0006
2.11	.4826	.0174	2.68	.4963	.0037	3.25	.4994	.0006
2.12	.4830	.0170	2.69	.4964	.0036	3.30	.4995	.0005
2.13	.4834	.0166	2.70	.4965	.0035	3.35	.4996	.0004
2.14	.4838	.0162	2.71	.4966	.0034	3.40	.4997	.0003
2.15	.4842	.0158	2.72	.4967	.0033	3.45	.4997	.0003
2.16	.4846	.0154	2.73	.4968	.0032	3.50	.4998	.0002
2.17	.4850	.0150	2.74	.4969	.0031	3.60	.4998	.0002
2.18	.4854	.0146	2.75	.4970	.0030	3.70	.4999	.0001
2.19	.4857	.0143	2.76	.4971	.0029	3.80	.4999	.0001
2.20	.4861	.0139	2.77	.4972	.0028	3.90	.49995	.00005
2.21	.4864	.0136	2.78	.4973	.0027	4.00	.49997	.00003

SOURCE: From *Basic Statistics Tales of Distributions, 6th edition*, by C. Spatz © 1997. Reprinted with permission of Wadsworth, a division of Thomson Learning. Fax 800 730-2215.

APPENDIX B   The *t* distribution.

df	α level for two-tailed test					
	.20	.10	.05	.02	.01	.001
	α level for one-tailed test					
	.10	.05	.025	.01	.005	.0005
1	3.078	6.314	12.706	31.821	63.657	636.619
2	1.886	2.920	4.303	6.965	9.925	31.598
3	1.638	2.353	3.182	4.541	5.841	12.924
4	1.533	2.132	2.776	3.747	4.604	8.610
5	1.476	2.015	2.571	3.365	4.032	6.869
6	1.440	1.943	2.447	3.143	3.707	5.959
7	1.415	1.895	2.365	2.998	3.499	5.408
8	1.397	1.860	2.306	2.896	3.355	5.041
9	1.383	1.833	2.262	2.821	3.250	4.781
10	1.372	1.812	2.228	2.764	3.169	4.587
11	1.363	1.796	2.201	2.718	3.106	4.437
12	1.356	1.782	2.179	2.681	3.055	4.318
13	1.350	1.771	2.160	2.650	3.012	4.221
14	1.345	1.761	2.145	2.624	2.977	4.140
15	1.341	1.753	2.131	2.602	2.947	4.073
16	1.337	1.746	2.120	2.583	2.921	4.015
17	1.333	1.740	2.110	2.567	2.898	3.965
18	1.330	1.734	2.101	2.552	2.878	3.922
19	1.328	1.729	2.093	2.539	2.861	3.883
20	1.325	1.725	2.086	2.528	2.845	3.850
21	1.323	1.721	2.080	2.518	2.831	3.819
22	1.321	1.717	2.074	2.508	2.819	3.792
23	1.319	1.714	2.069	2.500	2.807	3.767
24	1.318	1.711	2.064	2.492	2.797	3.745
25	1.316	1.708	2.060	2.485	2.787	3.725
26	1.315	1.706	2.056	2.479	2.779	3.707
27	1.314	1.703	2.052	2.474	2.771	3.690
28	1.313	1.701	2.048	2.467	2.763	3.674
29	1.311	1.699	2.045	2.462	2.756	3.659
30	1.310	1.697	2.042	2.457	2.750	3.646
40	1.303	1.684	2.021	2.423	2.704	3.551
60	1.296	1.671	2.000	2.390	2.660	3.460
120	1.289	1.658	1.980	2.358	2.617	3.373
∞	1.282	1.645	1.960	2.326	2.576	3.291

To be significant the *t* value obtained from the data must be equal to or greater than the value shown in the table.
SOURCE: From (c) 1963 R. A. Fisher and F. Yates, *Statistical tables for biological, agricultural, and medical research* (6th ed.). Reprinted by permission of Addison Wesley Longman Limited. Reprinted by permission of Pearson Education Limited.

APPENDIX C The F distribution.

α levels of .05 (lightface) and .01 (**boldface**) for the distribution of F

Each cell shows .05 value (lightface) / .01 value (**boldface**).

Degrees of Freedom (for the numerator of F ratio)

Denom. df	1	2	3	4	5	6	7	8	9	10	11	12	14	16	20	24	30	40	50	75	100	200	500	∞
1	161 / **4,052**	200 / **4,999**	216 / **5,403**	225 / **5,625**	230 / **5,764**	234 / **5,859**	237 / **5,928**	239 / **5,981**	241 / **6,022**	242 / **6,056**	243 / **6,082**	244 / **6,106**	245 / **6,142**	246 / **6,169**	248 / **6,208**	249 / **6,234**	250 / **6,258**	251 / **6,286**	252 / **6,302**	253 / **6,323**	253 / **6,334**	254 / **6,352**	254 / **6,361**	254 / **6,366**
2	18.51 / **98.49**	19.00 / **99.00**	19.16 / **99.17**	19.25 / **99.25**	19.30 / **99.30**	19.33 / **99.33**	19.36 / **99.34**	19.37 / **99.36**	19.38 / **99.39**	19.39 / **99.40**	19.40 / **99.41**	19.41 / **99.42**	19.42 / **99.43**	19.43 / **99.44**	19.44 / **99.45**	19.45 / **99.46**	19.46 / **99.47**	19.47 / **99.48**	19.47 / **99.49**	19.48 / **99.49**	19.49 / **99.49**	19.49 / **99.49**	19.50 / **99.50**	19.50 / **99.50**
3	10.13 / **34.12**	9.55 / **30.82**	9.28 / **29.46**	9.12 / **28.71**	9.01 / **28.24**	8.94 / **27.91**	8.88 / **27.67**	8.84 / **27.49**	8.81 / **27.34**	8.78 / **27.23**	8.76 / **27.13**	8.74 / **27.05**	8.71 / **26.92**	8.69 / **26.83**	8.66 / **26.69**	8.64 / **26.60**	8.62 / **26.50**	8.60 / **26.41**	8.58 / **26.35**	8.57 / **26.27**	8.56 / **26.23**	8.54 / **26.18**	8.54 / **26.14**	8.53 / **26.12**
4	7.71 / **21.20**	6.94 / **18.00**	6.59 / **16.69**	6.39 / **15.98**	6.26 / **15.52**	6.16 / **15.21**	6.09 / **14.98**	6.04 / **14.80**	6.00 / **14.66**	5.96 / **14.54**	5.93 / **14.45**	5.91 / **14.37**	5.87 / **14.24**	5.84 / **14.15**	5.80 / **14.02**	5.77 / **13.93**	5.74 / **13.83**	5.71 / **13.74**	5.70 / **13.69**	5.68 / **13.61**	5.66 / **13.57**	5.65 / **13.52**	5.64 / **13.48**	5.63 / **13.46**
5	6.61 / **16.26**	5.79 / **13.27**	5.41 / **12.06**	5.19 / **11.39**	5.05 / **10.97**	4.95 / **10.67**	4.88 / **10.45**	4.82 / **10.27**	4.78 / **10.15**	4.74 / **10.05**	4.70 / **9.96**	4.68 / **9.89**	4.64 / **9.77**	4.60 / **9.68**	4.56 / **9.55**	4.53 / **9.47**	4.50 / **9.38**	4.46 / **9.29**	4.44 / **9.24**	4.42 / **9.17**	4.40 / **9.13**	4.38 / **9.07**	4.37 / **9.04**	4.36 / **9.02**
6	5.99 / **13.74**	5.14 / **10.92**	4.76 / **9.78**	4.53 / **9.15**	4.39 / **8.75**	4.28 / **8.47**	4.21 / **8.26**	4.15 / **8.10**	4.10 / **7.98**	4.06 / **7.87**	4.03 / **7.79**	4.00 / **7.72**	3.96 / **7.60**	3.92 / **7.52**	3.87 / **7.39**	3.84 / **7.31**	3.81 / **7.23**	3.77 / **7.14**	3.75 / **7.09**	3.72 / **7.02**	3.71 / **6.99**	3.69 / **6.94**	3.68 / **6.90**	3.67 / **6.88**
7	5.59 / **12.25**	4.74 / **9.55**	4.35 / **8.45**	4.12 / **7.85**	3.97 / **7.46**	3.87 / **7.19**	3.79 / **7.00**	3.73 / **6.84**	3.68 / **6.71**	3.63 / **6.62**	3.60 / **6.54**	3.57 / **6.47**	3.52 / **6.35**	3.49 / **6.27**	3.44 / **6.15**	3.41 / **6.07**	3.38 / **5.98**	3.34 / **5.90**	3.32 / **5.85**	3.29 / **5.78**	3.28 / **5.75**	3.25 / **5.70**	3.24 / **5.67**	3.23 / **5.65**
8	5.32 / **11.26**	4.46 / **8.65**	4.07 / **7.59**	3.84 / **7.01**	3.69 / **6.63**	3.58 / **6.37**	3.50 / **6.19**	3.44 / **6.03**	3.39 / **5.91**	3.34 / **5.82**	3.31 / **5.74**	3.28 / **5.67**	3.23 / **5.56**	3.20 / **5.48**	3.15 / **5.36**	3.12 / **5.28**	3.08 / **5.20**	3.05 / **5.11**	3.03 / **5.06**	3.00 / **5.00**	2.98 / **4.96**	2.96 / **4.91**	2.94 / **4.88**	2.93 / **4.86**
9	5.12 / **10.56**	4.26 / **8.02**	3.86 / **6.99**	3.63 / **6.42**	3.48 / **6.06**	3.37 / **5.80**	3.29 / **5.62**	3.23 / **5.47**	3.18 / **5.35**	3.13 / **5.26**	3.10 / **5.18**	3.07 / **5.11**	3.02 / **5.00**	2.98 / **4.92**	2.93 / **4.80**	2.90 / **4.73**	2.86 / **4.64**	2.82 / **4.56**	2.80 / **4.51**	2.77 / **4.45**	2.76 / **4.41**	2.73 / **4.36**	2.72 / **4.33**	2.71 / **4.31**
10	4.96 / **10.04**	4.10 / **7.56**	3.71 / **6.55**	3.48 / **5.99**	3.33 / **5.64**	3.22 / **5.39**	3.14 / **5.21**	3.07 / **5.06**	3.02 / **4.95**	2.97 / **4.85**	2.94 / **4.78**	2.91 / **4.71**	2.86 / **4.60**	2.82 / **4.52**	2.77 / **4.41**	2.74 / **4.33**	2.70 / **4.25**	2.67 / **4.17**	2.64 / **4.12**	2.61 / **4.05**	2.59 / **4.01**	2.56 / **3.96**	2.55 / **3.93**	2.54 / **3.91**
11	4.84 / **9.65**	3.98 / **7.20**	3.59 / **6.22**	3.36 / **5.67**	3.20 / **5.32**	3.09 / **5.07**	3.01 / **4.88**	2.95 / **4.74**	2.90 / **4.63**	2.86 / **4.54**	2.82 / **4.46**	2.79 / **4.40**	2.74 / **4.29**	2.70 / **4.21**	2.65 / **4.10**	2.61 / **4.02**	2.57 / **3.94**	2.53 / **3.86**	2.50 / **3.80**	2.47 / **3.74**	2.45 / **3.70**	2.42 / **3.66**	2.41 / **3.62**	2.40 / **3.60**
12	4.75 / **9.33**	3.88 / **6.93**	3.49 / **5.95**	3.26 / **5.41**	3.11 / **5.06**	3.00 / **4.82**	2.92 / **4.65**	2.85 / **4.50**	2.80 / **4.39**	2.76 / **4.30**	2.72 / **4.22**	2.69 / **4.16**	2.64 / **4.05**	2.60 / **3.98**	2.54 / **3.86**	2.50 / **3.78**	2.46 / **3.70**	2.42 / **3.61**	2.40 / **3.56**	2.36 / **3.49**	2.35 / **3.46**	2.32 / **3.41**	2.31 / **3.38**	2.30 / **3.36**
13	4.67 / **9.07**	3.80 / **6.70**	3.41 / **5.74**	3.18 / **5.20**	3.02 / **4.86**	2.92 / **4.62**	2.84 / **4.44**	2.77 / **4.30**	2.72 / **4.19**	2.67 / **4.10**	2.63 / **4.02**	2.60 / **3.96**	2.55 / **3.85**	2.51 / **3.78**	2.46 / **3.67**	2.42 / **3.59**	2.38 / **3.51**	2.34 / **3.42**	2.32 / **3.37**	2.28 / **3.30**	2.26 / **3.27**	2.24 / **3.28**	2.22 / **3.18**	2.21 / **3.16**

Degrees of freedom (for the denominator of the F ratio)

continued

To be statistically significant the F obtained from the data must be equal to or greater than the value shown in the table.
SOURCE: From *Statistical Methods*, by G. W. Snedecor and W. W. Cochran, (7th ed.). Copyright © 1980 Iowa State University Press. Reprinted with permission.

APPENDIX C (continued)

Degrees of Freedom (for the numerator of F ratio)

Degrees of freedom (for the denominator of the F ratio)

Each cell gives the upper value (regular type) and the lower value (**bold type**).

df	1	2	3	4	5	6	7	8	9	10	11	12	14	16	20	24	30	40	50	75	100	200	500	∞
14	4.60 **8.86**	3.74 **6.51**	3.34 **5.56**	3.11 **5.03**	2.96 **4.69**	2.85 **4.46**	2.77 **4.28**	2.70 **4.14**	2.65 **4.03**	2.60 **3.94**	2.56 **3.86**	2.53 **3.80**	2.48 **3.70**	2.44 **3.62**	2.39 **3.51**	2.35 **3.43**	2.31 **3.34**	2.27 **3.26**	2.24 **3.21**	2.21 **3.14**	2.19 **3.11**	2.16 **3.06**	2.14 **3.02**	2.13 **3.00**
15	4.54 **8.68**	3.68 **6.36**	3.29 **5.52**	3.06 **4.89**	2.90 **4.56**	2.79 **4.32**	2.70 **4.14**	2.64 **4.00**	2.59 **3.89**	2.55 **3.80**	2.51 **3.73**	2.48 **3.67**	2.43 **3.56**	2.39 **3.48**	2.33 **3.36**	2.29 **3.29**	2.25 **3.20**	2.21 **3.12**	2.18 **3.07**	2.15 **3.00**	2.12 **2.97**	2.10 **2.92**	2.08 **2.89**	2.07 **2.87**
16	4.49 **8.53**	3.63 **6.23**	3.24 **5.29**	3.01 **4.77**	2.85 **4.44**	2.74 **4.20**	2.66 **4.03**	2.59 **3.89**	2.54 **3.78**	2.49 **3.69**	2.45 **3.61**	2.42 **3.55**	2.37 **3.45**	2.33 **3.37**	2.28 **3.25**	2.24 **3.18**	2.20 **3.10**	2.16 **3.01**	2.13 **2.96**	2.09 **2.98**	2.07 **2.86**	2.04 **2.80**	2.02 **2.77**	2.01 **2.75**
17	4.45 **8.40**	3.59 **6.11**	3.20 **5.18**	2.96 **4.67**	2.81 **4.34**	2.70 **4.10**	2.62 **3.93**	2.55 **3.79**	2.50 **3.68**	2.45 **3.59**	2.41 **3.52**	2.38 **3.45**	2.33 **3.35**	2.29 **3.27**	2.23 **3.16**	2.19 **3.08**	2.15 **3.00**	2.11 **2.92**	2.08 **2.86**	2.04 **2.79**	2.02 **2.76**	1.99 **2.70**	1.97 **2.67**	1.96 **2.65**
18	4.41 **8.28**	3.55 **6.01**	3.16 **5.09**	2.93 **4.58**	2.77 **4.25**	2.66 **4.01**	2.58 **3.85**	2.51 **3.71**	2.46 **3.60**	2.41 **3.51**	2.37 **3.44**	2.34 **3.37**	2.29 **3.27**	2.25 **3.19**	2.19 **3.07**	2.15 **3.00**	2.11 **2.91**	2.07 **2.83**	2.04 **2.78**	2.00 **2.71**	1.98 **2.68**	1.95 **2.62**	1.93 **2.59**	1.92 **2.57**
19	4.38 **8.18**	3.52 **5.93**	3.13 **5.01**	2.90 **4.50**	2.74 **4.17**	2.63 **3.94**	2.55 **3.77**	2.48 **3.63**	2.43 **3.52**	2.38 **3.43**	2.34 **3.36**	2.31 **3.30**	2.26 **3.19**	2.21 **3.12**	2.15 **3.00**	2.11 **2.92**	2.07 **2.84**	2.02 **2.76**	2.00 **2.70**	1.96 **2.63**	1.94 **2.60**	1.91 **2.54**	1.90 **2.51**	1.88 **2.49**
20	4.35 **8.10**	3.49 **5.85**	3.10 **4.94**	2.87 **4.43**	2.71 **4.10**	2.60 **3.87**	2.52 **3.71**	2.45 **3.56**	2.40 **3.45**	2.35 **3.37**	2.31 **3.30**	2.28 **3.23**	2.23 **3.13**	2.18 **3.05**	2.12 **2.94**	2.08 **2.86**	2.04 **2.77**	1.99 **2.69**	1.96 **2.63**	1.92 **2.56**	1.90 **2.53**	1.87 **2.47**	1.85 **2.44**	1.84 **2.42**
21	4.32 **8.02**	3.47 **5.78**	3.07 **4.87**	2.84 **4.37**	2.68 **4.04**	2.57 **3.81**	2.49 **3.65**	2.42 **3.51**	2.37 **3.40**	2.32 **3.31**	2.28 **3.24**	2.25 **3.17**	2.20 **3.07**	2.15 **2.99**	2.09 **2.88**	2.05 **2.80**	2.00 **2.72**	1.96 **2.63**	1.93 **2.58**	1.89 **2.51**	1.87 **2.47**	1.84 **2.42**	1.82 **2.38**	1.81 **2.36**
22	4.30 **7.94**	3.44 **5.72**	3.05 **4.82**	2.82 **4.31**	2.66 **3.99**	2.55 **3.76**	2.47 **3.59**	2.40 **3.45**	2.35 **3.35**	2.30 **3.26**	2.26 **3.18**	2.23 **3.12**	2.18 **3.02**	2.13 **2.94**	2.07 **2.83**	2.03 **2.75**	1.98 **2.67**	1.93 **2.58**	1.91 **2.53**	1.87 **2.46**	1.84 **2.42**	1.81 **2.37**	1.80 **2.33**	1.78 **2.31**
23	4.28 **7.88**	3.42 **5.66**	3.03 **4.76**	2.80 **4.26**	2.64 **3.94**	2.53 **3.71**	2.45 **3.54**	2.38 **3.41**	2.32 **3.30**	2.28 **3.21**	2.24 **3.14**	2.20 **3.07**	2.14 **2.97**	2.10 **2.89**	2.04 **2.78**	2.00 **2.70**	1.96 **2.62**	1.91 **2.53**	1.88 **2.48**	1.84 **2.41**	1.82 **2.37**	1.79 **2.32**	1.77 **2.28**	1.76 **2.26**
24	4.26 **7.82**	3.40 **5.61**	3.01 **4.72**	2.78 **4.22**	2.62 **3.90**	2.51 **3.67**	2.43 **3.50**	2.36 **3.36**	2.30 **3.25**	2.26 **3.17**	2.22 **3.09**	2.18 **3.03**	2.13 **2.93**	2.09 **2.85**	2.02 **2.74**	1.98 **2.66**	1.94 **2.58**	1.89 **2.49**	1.86 **2.44**	1.82 **2.36**	1.80 **2.33**	1.76 **2.27**	1.74 **2.23**	1.73 **2.21**
25	4.24 **7.77**	3.38 **5.57**	2.99 **4.68**	2.76 **4.18**	2.60 **3.86**	2.49 **3.63**	2.41 **3.46**	2.34 **3.32**	2.28 **3.21**	2.24 **3.13**	2.20 **3.05**	2.16 **2.99**	2.11 **2.89**	2.06 **2.81**	2.00 **2.70**	1.96 **2.62**	1.92 **2.54**	1.87 **2.45**	1.84 **2.40**	1.80 **2.32**	1.77 **2.29**	1.74 **2.23**	1.72 **2.19**	1.71 **2.17**
26	4.22 **7.72**	3.37 **5.53**	2.98 **4.64**	2.74 **4.14**	2.59 **3.82**	2.47 **3.59**	2.39 **3.42**	2.32 **3.29**	2.27 **3.17**	2.22 **3.09**	2.18 **3.02**	2.15 **2.96**	2.10 **2.86**	2.05 **2.77**	1.99 **2.66**	1.95 **2.58**	1.90 **2.50**	1.85 **2.41**	1.82 **2.36**	1.78 **2.28**	1.76 **2.25**	1.72 **2.19**	1.70 **2.15**	1.69 **2.13**

continued

APPENDIX C (continued)

Degrees of freedom (for the numerator of F ratio)

den. df	1	2	3	4	5	6	7	8	9	10	11	12	14	16	20	24	30	40	50	75	100	200	500	∞
27	4.21 **7.68**	3.35 **5.49**	2.96 **4.60**	2.73 **4.11**	2.57 **3.79**	2.46 **3.56**	2.37 **3.39**	2.30 **3.26**	2.25 **3.14**	2.20 **3.06**	2.16 **2.98**	2.13 **2.93**	2.08 **2.83**	2.03 **2.74**	1.97 **2.63**	1.93 **2.55**	1.88 **2.47**	1.84 **2.38**	1.80 **2.33**	1.76 **2.25**	1.74 **2.21**	1.71 **2.16**	1.68 **2.12**	1.67 **2.10**
28	4.20 **7.64**	3.34 **5.45**	2.95 **4.57**	2.71 **4.07**	2.56 **3.76**	2.44 **3.53**	2.36 **3.36**	2.29 **3.23**	2.24 **3.11**	2.19 **3.03**	2.15 **2.95**	2.12 **2.90**	2.06 **2.80**	2.02 **2.71**	1.96 **2.60**	1.91 **2.52**	1.87 **2.44**	1.81 **2.35**	1.78 **2.30**	1.75 **2.22**	1.72 **2.18**	1.69 **2.13**	1.67 **2.09**	1.65 **2.06**
29	4.18 **7.60**	3.33 **5.42**	2.93 **4.54**	2.70 **4.04**	2.54 **3.73**	2.43 **3.50**	2.35 **3.33**	2.28 **3.20**	2.22 **3.08**	2.18 **3.00**	2.14 **2.92**	2.10 **2.87**	2.05 **2.77**	2.00 **2.68**	1.94 **2.57**	1.90 **2.49**	1.85 **2.41**	1.80 **2.32**	1.77 **2.27**	1.73 **2.19**	1.71 **2.15**	1.68 **2.10**	1.65 **2.06**	1.64 **2.03**
30	4.17 **7.56**	3.32 **5.39**	2.92 **4.51**	2.69 **4.02**	2.53 **3.70**	2.42 **3.47**	2.34 **3.30**	2.27 **3.17**	2.21 **3.06**	2.16 **2.98**	2.12 **2.90**	2.09 **2.84**	2.04 **2.74**	1.99 **2.66**	1.93 **2.55**	1.89 **2.47**	1.84 **2.38**	1.79 **2.29**	1.76 **2.24**	1.72 **2.16**	1.69 **2.13**	1.66 **2.07**	1.64 **2.03**	1.62 **2.01**
32	4.15 **7.50**	3.30 **5.34**	2.90 **4.46**	2.67 **3.97**	2.51 **3.66**	2.40 **3.42**	2.32 **3.25**	2.25 **3.12**	2.19 **3.01**	2.14 **2.94**	2.10 **2.86**	2.07 **2.80**	2.02 **2.70**	1.97 **2.62**	1.91 **2.51**	1.86 **2.42**	1.82 **2.34**	1.76 **2.25**	1.74 **2.20**	1.69 **2.12**	1.67 **2.08**	1.64 **2.02**	1.61 **1.98**	1.59 **1.96**
34	4.13 **7.44**	3.28 **5.29**	2.88 **4.42**	2.65 **3.93**	2.49 **3.61**	2.38 **3.38**	2.30 **3.21**	2.23 **3.08**	2.17 **2.97**	2.12 **2.89**	2.08 **2.82**	2.05 **2.76**	2.00 **2.66**	1.95 **2.58**	1.89 **2.47**	1.84 **2.38**	1.80 **2.30**	1.74 **2.21**	1.71 **2.15**	1.67 **2.08**	1.64 **2.04**	1.61 **1.98**	1.59 **1.94**	1.57 **1.91**
36	4.11 **7.39**	3.26 **5.24**	2.86 **4.38**	2.63 **3.89**	2.48 **3.58**	2.36 **3.35**	2.28 **3.18**	2.21 **3.04**	2.15 **2.94**	2.10 **2.86**	2.06 **2.78**	2.03 **2.72**	1.98 **2.62**	1.93 **2.54**	1.87 **2.43**	1.82 **2.35**	1.78 **2.26**	1.72 **2.17**	1.69 **2.12**	1.65 **2.04**	1.62 **2.00**	1.59 **1.94**	1.56 **1.90**	1.55 **1.87**
38	4.10 **7.35**	3.25 **5.21**	2.85 **4.34**	2.62 **3.86**	2.46 **3.54**	2.35 **3.32**	2.26 **3.15**	2.19 **3.02**	2.14 **2.91**	2.09 **2.82**	2.05 **2.75**	2.02 **2.69**	1.96 **2.59**	1.92 **2.51**	1.85 **2.40**	1.80 **2.32**	1.76 **2.22**	1.71 **2.14**	1.67 **2.08**	1.63 **2.00**	1.60 **1.97**	1.57 **1.90**	1.54 **1.86**	1.53 **1.84**
40	4.08 **7.31**	3.23 **5.18**	2.84 **4.31**	2.61 **3.83**	2.45 **3.51**	2.34 **3.29**	2.25 **3.12**	2.18 **2.99**	2.12 **2.88**	2.07 **2.80**	2.04 **2.73**	2.00 **2.66**	1.95 **2.56**	1.90 **2.49**	1.84 **2.37**	1.79 **2.29**	1.74 **2.20**	1.69 **2.11**	1.66 **2.05**	1.61 **1.97**	1.59 **1.94**	1.55 **1.88**	1.53 **1.84**	1.51 **1.81**
42	4.07 **7.27**	3.22 **5.15**	2.83 **4.29**	2.59 **3.80**	2.44 **3.49**	2.32 **3.26**	2.24 **3.10**	2.17 **2.96**	2.11 **2.86**	2.06 **2.77**	2.02 **2.70**	1.99 **2.64**	1.94 **2.54**	1.89 **2.46**	1.82 **2.35**	1.78 **2.26**	1.73 **2.17**	1.68 **2.08**	1.64 **2.02**	1.60 **1.94**	1.57 **1.91**	1.54 **1.85**	1.51 **1.80**	1.49 **1.78**
44	4.06 **7.25**	3.21 **5.12**	2.82 **4.26**	2.58 **3.78**	2.43 **3.46**	2.31 **3.24**	2.23 **3.07**	2.16 **2.94**	2.10 **2.84**	2.05 **2.75**	2.01 **2.68**	1.98 **2.62**	1.92 **2.52**	1.88 **2.44**	1.81 **2.32**	1.76 **2.24**	1.72 **2.15**	1.66 **2.06**	1.63 **2.00**	1.58 **1.92**	1.56 **1.88**	1.52 **1.82**	1.50 **1.78**	1.48 **1.75**
46	4.05 **7.21**	3.20 **5.10**	2.81 **4.24**	2.57 **3.76**	2.42 **3.44**	2.30 **3.22**	2.22 **3.05**	2.14 **2.92**	2.09 **2.82**	2.04 **2.73**	2.00 **2.66**	1.97 **2.60**	1.91 **2.50**	1.87 **2.42**	1.80 **2.30**	1.75 **2.22**	1.71 **2.13**	1.65 **2.04**	1.62 **1.98**	1.57 **1.90**	1.54 **1.86**	1.51 **1.80**	1.48 **1.76**	1.46 **1.72**
48	4.04 **7.19**	3.19 **5.08**	2.80 **4.22**	2.56 **3.74**	2.41 **3.42**	2.30 **3.20**	2.21 **3.04**	2.14 **2.90**	2.08 **2.80**	2.03 **2.71**	1.99 **2.64**	1.96 **2.58**	1.90 **2.48**	1.86 **2.40**	1.79 **2.28**	1.74 **2.20**	1.70 **2.11**	1.64 **2.02**	1.61 **1.96**	1.56 **1.88**	1.53 **1.84**	1.50 **1.78**	1.47 **1.73**	1.45 **1.70**

Degrees of freedom (for the denominator of the F ratio)

continued

APPENDIX C (*continued*)

Degrees of freedom (for the numerator of F ratio)

Degrees of freedom (for the denominator of the F ratio)

df (denom)	1	2	3	4	5	6	7	8	9	10	11	12	14	16	20	24	30	40	50	75	100	200	500	∞
50	4.03 / 7.17	3.18 / 5.06	2.79 / 4.20	2.56 / 3.72	2.40 / 3.41	2.29 / 3.18	2.20 / 3.02	2.13 / 2.88	2.07 / 2.78	2.02 / 2.70	1.98 / 2.62	1.95 / 2.56	1.90 / 2.46	1.85 / 2.39	1.78 / 2.26	1.74 / 2.18	1.69 / 2.10	1.63 / 2.00	1.60 / 1.94	1.55 / 1.86	1.52 / 1.82	1.48 / 1.76	1.46 / 1.71	1.44 / 1.68
55	4.02 / 7.12	3.17 / 5.01	2.78 / 4.16	2.54 / 3.68	2.38 / 3.37	2.27 / 3.15	2.18 / 2.98	2.11 / 2.85	2.05 / 2.75	2.00 / 2.66	1.97 / 2.59	1.93 / 2.53	1.88 / 2.43	1.83 / 2.35	1.76 / 2.23	1.72 / 2.15	1.67 / 2.06	1.61 / 1.96	1.58 / 1.90	1.52 / 1.82	1.50 / 1.78	1.46 / 1.71	1.43 / 1.66	1.41 / 1.64
60	4.00 / 7.08	3.15 / 4.98	2.76 / 4.13	2.52 / 3.65	2.37 / 3.34	2.25 / 3.12	2.17 / 2.95	2.10 / 2.82	2.04 / 2.72	1.99 / 2.63	1.95 / 2.56	1.92 / 2.50	1.86 / 2.40	1.81 / 2.32	1.75 / 2.20	1.70 / 2.12	1.65 / 2.03	1.59 / 1.93	1.56 / 1.87	1.50 / 1.79	1.48 / 1.74	1.44 / 1.68	1.41 / 1.63	1.39 / 1.60
65	3.99 / 7.04	3.14 / 4.95	2.75 / 4.10	2.51 / 3.62	2.36 / 3.31	2.24 / 3.09	2.15 / 2.93	2.08 / 2.79	2.02 / 2.70	1.98 / 2.61	1.94 / 2.54	1.90 / 2.47	1.85 / 2.37	1.80 / 2.30	1.73 / 2.18	1.68 / 2.09	1.63 / 2.00	1.57 / 1.90	1.54 / 1.84	1.49 / 1.76	1.46 / 1.71	1.42 / 1.64	1.39 / 1.60	1.37 / 1.56
70	3.98 / 7.01	3.13 / 4.92	2.74 / 4.08	2.50 / 3.60	2.35 / 3.29	2.23 / 3.07	2.14 / 2.91	2.07 / 2.77	2.01 / 2.67	1.97 / 2.59	1.93 / 2.51	1.89 / 2.45	1.84 / 2.35	1.79 / 2.28	1.72 / 2.15	1.67 / 2.07	1.62 / 1.98	1.56 / 1.88	1.53 / 1.82	1.47 / 1.74	1.45 / 1.69	1.40 / 1.62	1.37 / 1.56	1.35 / 1.53
80	3.96 / 6.96	3.11 / 4.88	2.72 / 4.04	2.48 / 3.56	2.33 / 3.25	2.21 / 3.04	2.12 / 2.87	2.05 / 2.74	1.99 / 2.64	1.95 / 2.55	1.91 / 2.48	1.88 / 2.41	1.82 / 2.32	1.77 / 2.24	1.70 / 2.11	1.65 / 2.03	1.60 / 1.94	1.54 / 1.84	1.51 / 1.78	1.45 / 1.70	1.42 / 1.65	1.38 / 1.57	1.35 / 1.52	1.32 / 1.49
100	3.94 / 6.90	3.09 / 4.82	2.70 / 3.98	2.46 / 3.51	2.30 / 3.20	2.19 / 2.99	2.10 / 2.82	2.03 / 2.69	1.97 / 2.59	1.92 / 2.51	1.88 / 2.43	1.85 / 2.36	1.79 / 2.26	1.75 / 2.19	1.68 / 2.06	1.63 / 1.98	1.57 / 1.89	1.51 / 1.79	1.48 / 1.73	1.42 / 1.64	1.39 / 1.59	1.34 / 1.51	1.30 / 1.46	1.28 / 1.43
125	3.92 / 6.84	3.07 / 4.78	2.68 / 3.94	2.44 / 3.47	2.29 / 3.17	2.17 / 2.95	2.08 / 2.79	2.01 / 2.65	1.95 / 2.56	1.90 / 2.47	1.86 / 2.40	1.83 / 2.33	1.77 / 2.23	1.72 / 2.15	1.65 / 2.03	1.60 / 1.94	1.55 / 1.85	1.49 / 1.75	1.45 / 1.68	1.39 / 1.59	1.36 / 1.54	1.31 / 1.46	1.27 / 1.40	1.25 / 1.37
150	3.91 / 6.81	3.06 / 4.75	2.67 / 3.91	2.43 / 3.44	2.27 / 3.14	2.16 / 2.92	2.07 / 2.76	2.00 / 2.62	1.94 / 2.53	1.89 / 2.44	1.85 / 2.37	1.82 / 2.30	1.76 / 2.20	1.71 / 2.12	1.64 / 2.00	1.59 / 1.91	1.54 / 1.83	1.47 / 1.72	1.44 / 1.66	1.37 / 1.56	1.34 / 1.51	1.29 / 1.43	1.25 / 1.37	1.22 / 1.33
200	3.89 / 6.76	3.04 / 4.71	2.65 / 3.88	2.41 / 3.41	2.26 / 3.11	2.14 / 2.90	2.05 / 2.73	1.98 / 2.60	1.92 / 2.50	1.87 / 2.41	1.83 / 2.34	1.80 / 2.28	1.74 / 2.17	1.69 / 2.09	1.62 / 1.97	1.57 / 1.88	1.52 / 1.79	1.45 / 1.69	1.42 / 1.62	1.35 / 1.53	1.32 / 1.48	1.26 / 1.39	1.22 / 1.33	1.19 / 1.28
400	3.86 / 6.70	3.02 / 4.66	2.62 / 3.83	2.39 / 3.36	2.23 / 3.06	2.12 / 2.85	2.03 / 2.69	1.96 / 2.55	1.90 / 2.46	1.85 / 2.37	1.81 / 2.29	1.78 / 2.23	1.72 / 2.12	1.67 / 2.04	1.60 / 1.92	1.54 / 1.84	1.49 / 1.74	1.42 / 1.64	1.38 / 1.57	1.32 / 1.47	1.28 / 1.42	1.22 / 1.32	1.16 / 1.24	1.13 / 1.19
1000	3.85 / 6.66	3.00 / 4.62	2.61 / 3.80	2.38 / 3.34	2.22 / 3.04	2.10 / 2.82	2.02 / 2.66	1.95 / 2.53	1.89 / 2.43	1.84 / 2.34	1.80 / 2.26	1.76 / 2.20	1.70 / 2.09	1.65 / 2.01	1.58 / 1.89	1.53 / 1.81	1.47 / 1.71	1.41 / 1.61	1.36 / 1.54	1.30 / 1.44	1.26 / 1.38	1.19 / 1.28	1.13 / 1.19	1.08 / 1.11
∞	3.84 / 6.63	2.99 / 4.60	2.60 / 3.78	2.37 / 3.32	2.21 / 3.02	2.09 / 2.80	2.01 / 2.64	1.94 / 2.51	1.88 / 2.41	1.83 / 2.32	1.79 / 2.24	1.75 / 2.18	1.69 / 2.07	1.64 / 1.99	1.57 / 1.87	1.52 / 1.79	1.46 / 1.69	1.40 / 1.59	1.35 / 1.52	1.28 / 1.41	1.24 / 1.36	1.17 / 1.25	1.11 / 1.15	1.00 / 1.00

APPENDIX D Critical values of the studentized range statistic (for Tukey HSD tests).

$\alpha = .05$

df error	Number of levels of the independent variable													
	2	3	4	5	6	7	8	9	10	11	12	13	14	15
1	17.97	26.98	2.82	37.07	40.41	43.12	45.40	47.36	49.07	50.59	51.96	53.20	54.33	55.36
2	6.08	8.33	9.80	10.88	11.74	12.44	13.03	13.54	13.99	14.39	14.75	15.08	15.38	15.65
3	4.50	5.91	6.82	7.50	8.04	8.48	8.85	9.18	9.46	9.72	9.95	10.15	10.35	10.53
4	3.93	5.04	5.76	6.29	6.71	7.05	7.35	7.60	7.33	8.03	8.21	8.37	8.52	8.66
5	3.64	4.60	5.22	5.67	6.03	6.33	6.58	6.80	7.00	7.17	7.32	7.47	7.60	7.72
6	3.46	4.34	4.90	5.31	5.63	5.90	6.12	6.32	6.49	6.65	6.79	6.92	7.03	7.14
7	3.34	4.16	4.68	5.06	5.36	5.61	5.82	6.00	6.16	6.30	6.43	6.55	6.66	6.76
8	3.26	4.04	4.53	4.89	5.17	5.40	5.60	5.77	5.92	6.05	6.18	6.29	6.39	6.48
9	3.20	3.95	4.42	4.76	5.02	5.24	5.43	5.60	5.74	5.87	5.98	6.09	6.19	6.28
10	3.15	3.88	4.33	4.65	4.91	5.12	5.30	5.46	5.60	5.72	5.83	5.94	6.03	6.11
11	3.11	3.82	4.26	4.57	4.82	5.03	5.20	5.35	5.49	5.60	5.71	5.81	5.90	5.98
12	3.08	3.77	4.20	4.51	4.75	4.95	5.12	5.26	5.40	5.51	5.62	5.71	5.79	5.88
13	3.06	3.74	4.15	4.45	4.69	4.88	5.05	5.19	5.32	5.43	5.53	5.63	5.71	5.79
14	3.03	3.70	4.11	4.41	4.64	4.83	4.99	5.13	5.25	5.36	5.46	5.55	5.64	5.71
15	3.01	3.67	4.08	4.37	4.60	4.78	4.94	5.08	5.20	5.31	5.40	5.49	5.57	5.65
16	3.00	3.65	4.05	4.33	4.56	4.74	4.90	5.03	5.15	5.26	5.35	5.44	5.52	5.59
17	2.98	3.63	4.02	4.30	4.52	4.70	4.86	4.99	5.11	5.21	5.31	5.39	5.47	5.54
18	2.97	3.61	4.00	4.28	4.50	4.67	4.82	4.96	5.07	5.17	5.27	5.35	5.43	5.50
19	2.96	3.59	3.98	4.25	4.47	4.64	4.79	4.92	5.04	5.14	5.23	5.32	5.39	5.46
20	2.95	3.58	3.96	4.23	4.44	4.62	4.77	4.90	5.01	5.11	5.20	5.28	5.36	5.43
24	2.92	3.53	3.90	4.17	4.37	4.54	4.68	4.81	4.92	5.01	5.10	5.18	5.25	5.32
30	2.89	3.49	3.84	4.10	4.30	4.46	4.60	4.72	4.82	4.92	5.00	5.08	5.15	5.21
40	2.86	3.44	3.79	4.04	4.23	4.39	4.52	4.64	4.74	4.82	4.90	4.98	5.04	5.11
60	2.83	3.40	3.74	3.98	4.16	4.31	4.44	4.55	4.65	4.73	4.81	4.88	4.94	5.00
120	2.80	3.36	3.69	3.92	4.10	4.24	4.36	4.47	4.56	4.64	4.71	4.78	4.84	4.90
∞	2.77	3.31	3.63	3.86	4.03	4.17	4.29	4.39	4.47	4.55	4.62	4.68	4.74	4.80

SOURCE: From "Tables of Range and Studentized Range," by M. L. Harter, *Annals of Mathematical Statistics, 31*, 1122-1147 (1960). Reprinted with permission.

APPENDIX D (continued).

$\alpha = .01$

Number of levels of the independent variable

df error	2	3	4	5	6	7	8	9	10	11	12	13	14	15
1	90.03	135.00	164.30	185.60	202.20	215.80	227.20	237.00	245.60	253.20	260.00	266.20	271.80	277.00
2	14.04	19.02	22.29	24.72	26.63	28.20	29.53	30.68	31.69	32.59	33.40	34.13	34.81	35.43
3	8.26	10.62	12.17	13.33	14.24	15.00	15.64	12.60	16.69	17.13	17.53	17.89	18.22	18.52
4	6.51	8.12	9.17	9.96	10.58	11.10	11.55	11.93	12.27	12.57	12.84	13.09	13.32	13.53
5	5.70	6.98	7.80	8.42	8.91	9.32	9.67	9.97	10.24	10.48	10.70	10.89	11.08	11.24
6	5.24	6.33	7.03	7.56	7.97	8.32	8.62	8.87	9.10	9.30	9.48	9.65	9.81	9.95
7	4.95	5.92	6.54	7.00	7.37	7.68	7.94	8.17	8.37	8.55	8.71	8.86	9.00	9.12
8	4.75	5.64	6.20	6.62	6.96	7.24	7.47	7.68	7.86	8.03	8.18	8.31	8.44	8.55
9	4.60	5.43	5.96	6.35	6.66	6.92	7.13	7.32	7.50	7.65	7.78	7.91	8.02	8.13
10	4.48	5.27	5.77	6.14	6.43	6.67	6.88	7.06	7.21	7.36	7.48	7.60	7.71	7.81
11	4.39	5.15	5.62	5.97	6.25	6.48	6.67	6.84	6.99	7.13	7.25	7.36	7.46	7.56
12	4.32	5.05	5.50	5.84	6.10	6.32	6.51	6.67	6.81	6.94	7.06	7.17	7.26	7.36
13	4.26	4.96	5.40	5.73	5.98	6.19	6.37	6.53	6.67	6.79	6.90	7.01	7.10	7.19
14	4.21	4.90	5.32	5.63	5.88	6.08	6.26	6.41	6.54	6.66	6.77	6.87	6.96	7.05
15	4.17	4.84	5.25	5.56	5.80	5.99	6.16	6.31	6.44	6.56	6.66	6.76	6.84	6.93
16	4.13	4.79	5.19	5.49	5.72	5.92	6.08	6.22	6.35	6.46	6.56	6.66	6.74	6.82
17	4.10	4.74	5.14	5.43	5.66	5.85	6.01	6.15	6.27	6.38	6.48	6.57	6.66	6.73
18	4.07	4.70	5.09	5.38	5.60	5.79	5.94	6.08	6.20	6.31	6.41	6.50	6.58	6.66
19	4.05	4.67	5.05	5.33	5.55	5.74	5.89	6.02	6.14	6.25	6.34	6.43	6.51	6.58
20	4.02	4.64	5.02	5.29	5.51	5.69	5.84	5.97	6.09	6.19	6.28	6.37	6.45	6.52
24	3.96	4.55	4.91	5.17	5.37	5.54	5.69	5.81	5.92	6.02	6.11	6.19	6.26	6.33
30	3.89	4.46	4.80	5.05	5.24	5.40	5.54	5.65	5.76	5.85	5.93	6.01	6.08	6.14
40	3.82	4.37	4.70	4.93	5.11	5.26	5.39	5.50	5.60	5.69	5.76	5.84	5.90	5.96
60	3.76	4.28	4.60	4.82	4.99	5.13	5.25	5.36	5.45	5.53	5.60	5.67	5.73	5.78
120	3.70	4.20	3.50	4.71	4.87	5.01	5.12	5.21	5.30	5.38	5.44	5.51	5.56	5.61
∞	3.64	4.12	4.40	4.60	4.76	4.88	4.99	5.08	5.16	5.23	5.29	5.35	5.40	5.45

REFERENCES

Aiken, L. S., & West, S. G. (1991). *Multiple regression: Testing and interpreting interactions.* Newbury Park, CA: Sage.

Berliner, D. C., & Biddle, B. J. (1995). *The manufactured crisis: Myths, fraud, and the attack on America's public schools.* New York: Addison-Wesley.

Berry, W. D., & Feldman, S. (1985). *Multiple regression in practice.* Newbury Park, CA: Sage.

Bracey, G. W. (1991). Why can't they be like we were? *Phi Delta Kappan* (October), 104-117.

Burger, J. M. (1987). Increased performance with increased personal control: A self-presentation interpretation. *Journal of Experimental Social Psychology, 23,* 350-360.

Cohen, J., & Cohen, P. (1975). *Applied multiple regression/correlation analysis for the behavioral sciences.* Hillsdale, NJ: Lawrence Erlbaum Associates.

Glass, G. V., & Hopkins, K. D. (1996). *Statistical methods in education and psychology* (3rd ed.). Boston: Allyn & Bacon.

Hinkle, D. E., Wiersma, W., & Jurs, S. G. (1998). *Applied statistics for the behavioral sciences* (4th Ed.). Boston: Houghton Mifflin.

Iverson, G. R., & Norpoth, H. (1987). *Analysis of variance* (2nd ed.) Newbury Park, CA: Sage.

Jaccard, J., Turrisi, R., & Wan, C. K. (1990). *Interaction effects in multiple regression.* Newbury Park, CA: Sage.

Marascuilo, L. A., & Serlin, R. C. (1988). *Statistical methods for the social and behavioral sciences* (pp. 472-516). New York: Freeman.

Marascuilo, L. A., & Serlin, R. C. (1988). *Statistical methods for the social and behavioral sciences.* New York: Freeman.

Midgley, C., Kaplan, A., Middleton, M., Maehr, M. L., Urdan, T., Anderman, L. H., Anderman, E., & Roeser, R. (1998). The development and validation of scales assessing students' achievement goal orientations. *Contemporary Educational Psychology, 23,* 113-131.

Mohr, L. B. (1990). *Understanding significance testing.* Newbury Park, CA: Sage.

Naglieri (1996). *The Naglieri nonverbal ability test.* San Antonio, TX: Harcourt Brace.

Pedhazur, E. J. (1982). *Multiple regression in behavioral research: Explanation and prediction* (2nd ed.). New York: Harcourt Brace.

Spatz, C. (2001) *Basic statistics: Tales of distributions* (7th ed.). Belmont, CA: Wadsworth.

Wildt, A. R., & Ahtola, O. T. (1978). *Analysis of covariance.* Newbury Park, CA: Sage.

INDEX OF TERMS

NOTE: The definitions for each of these terms appear at the end of the chapters in which they appear. In this index, the page number and chapter where the term first appears is provided. Some terms appear in more than one chapter. In those cases, multiple page and chapter numbers are provided.

GLOSSARY OF SYMBOLS

Σ	The sum of; to sum.
X	An individual, or raw, score in a distribution.
ΣX	The sum of X; adding up all of the scores in a distribution.
$\overline{X}$	The mean of a sample.
μ	The mean of a population.
n	The number of cases, or scores, in a sample.
N	The number of cases, or scores, in a population.
s^2	The sample variance.
s	The sample standard deviation.
σ	The population standard deviation.
σ^2	The population variance.
SS	The sum of squares, or sum of squared deviations.
z	A standard score.
$s_{\overline{x}}$	The standard error of the mean estimated from the sample standard deviation (i.e., when the population standard deviation is unknown).
$\sigma_{\overline{x}}$	The standard error of the mean when the population standard deviation is known.
p	p value, or probability.
α	Alpha level.
d	Effect size.
S	The standard deviation used in the effect size formula.
∞	Infinity.
H_0	The null hypothesis.
H_A or H_1	Symbols for the alternative hypothesis.
r	The sample Pearson correlation coefficient.
ρ	Rho, the population correlation coefficient.
s_r	The standard error of the correlation coefficient.
r^2	The coefficient of determination.
df	Degrees of freedom.
Φ	The phi coefficient, which is the correlation between two dichotomous variables.
$s_{\overline{x}_1-\overline{x}_2}$	The standard error of difference between two independent sample means.
$s_{\overline{D}}$	The standard error of the difference between to dependent, matched, or paired samples.
t	The t value.
MS_w	The mean square within groups.

MS_e	The mean square error (which is the same as the mean square within groups).
MS_b	The mean square between groups.
SS_e	The sum of squares error (or within groups).
SS_b	The sum of squares between groups.
SS_T	The sum of squares total.
X_T	The grand mean.
F	The F value.
K	The number of groups.
N	The number of cases in all of the groups combined.
n	The number of cases in a given group (for calculating SS_b).
N_T	The number of cases in each group (for Tukey HSD test).
$MS_{s \times T}$	The mean square for the interaction of subject by trial.
MS_T	The mean square for the differences between the trials.
$\hat{Y}$	The predicted value of Y, the dependent variable.
Y	The observed value of Y, the dependent variable.
b	The unstandardized regression coefficient.
a	The intercept.
e	The error term.
R	The multiple correlation coefficient.
R^2	The percentage of variance explained by the regression model.